Sooyong Park is an award-winning documentary filmmaker who has devoted over twenty years to studying and filming Siberian tigers. His groundbreaking tiger research was the subject of the film *Siberian Tiger Quest*. When not in Siberia, he lives in Seoul, Korea.

Based on his philosophy that "nature is to be observed, not directed", his work involves unimaginable patience and a painstaking devotion to becoming part of nature. His profound insights into the natural world and his warm affection for living creatures reveal that animals, like humans, have souls.

Praise for *The Great Soul of Siberia*:

"A wonderful evocation of the land and the habits of the desperately endangered Siberian tiger." *Independent*

"The year's best wildlife book could already have arrived."
Wanderlust

"Subtly intense... Park has a deep sense of oneness with the world around him. His close engagement with the forest ecology is the most extraordinary element of this remarkable book."
New Statesman

"It's a masterpiece. One of the most moving outdoor texts I've read in years. This is a book about love—one exceptional human being's love for the wild, beautiful and persecuted creatures to which his life is dedicated. It also comprehends a fortitude and hardihood so far beyond the everyday I was left shaking my head in astonished admiration." *The Great Outdoors*

"Sooyong's magical prose led me into little-known and breathtakingly beautiful forests, exposed me to the bitter cold of long winter months, and revealed the secret life of that most mysterious of cats, the Siberian tiger." JANE GOODALL

"A triumph of nature writing; an original and authentic voice from the wilderness." JONATHAN and ANGELA SCOTT

"The book is a love letter... To read it is to hear the voice of a remarkable man." *Daily Telegraph*

"Spellbinding... Park's book burns an indelible memory."
Country Life

THE GREAT SOUL OF SIBERIA

THE GREAT SOUL OF SIBERIA

In Search of the Elusive Siberian Tiger

SOOYONG PARK

William Collins
An imprint of HarperCollins*Publishers*
1 London Bridge Street
London SE1 9GF

WilliamCollinsBooks.com

This William Collins paperback edition published in 2017

22 21 20 19 18 17
10 9 8 7 6 5 4 3 2 1

First published in the United Kingdom by William Collins in 2016

First published in Canada by Greystone Books Ltd in 2015 as *Great Soul of Siberia: Passion Obsession, and One Man's Quest for the World's Most Elusive Tiger*

The Great Soul of Siberia is published under the support of the Literature Translation Institute of Korea (LTI Korea).

A catalogue record for this book is available from the British Library

ISBN 978-0-00-815617-6

Editing by Jennifer Croll
Copy-editing by Lana Okerlund
Text design by Nayeli Jimenez

Thank you to Vladimir Medvegev and Galina Salkina for providing some of the photographs, and to Sora Kim-Russell, who read and offered feedback on the translation.

Printed and bound in Great Britain by Clays Ltd, St Ives plc.

CONTENTS

FOREWORD

by John Vaillant

THIS BOOK IS a personal account of one of the most extraordinary wildlife studies ever undertaken. That its focus was on Siberian tigers, arguably the most difficult tiger subspecies to study in the wild, only makes Sooyong Park's accomplishment that much more remarkable. Many of his observations were made from an earthen bunker scarcely larger than a grave where, over a period of years, he sequestered himself for months at a time in subzero temperatures. The fact that Park survived these extended periods of solitary confinement with body and soul intact is astonishing in itself. That he emerged with enough film footage for seven documentaries and the inspiration for this beautiful, necessary book makes it a unique achievement.

Park's radical method was based on his understanding that Siberian tigers do not forgive disturbances to their natural order. If something doesn't look, smell, or feel right, the tiger will either find and destroy the cause or vacate the premises so thoroughly that any hope of future sightings there is lost. Park solved this problem by disappearing himself to the point that he would not register with the tigers' acute hearing, vision, or sense of smell. And then he waited—sometimes for months—for a tiger to come by. Relegating oneself to such a state of egoless oblivion is difficult, even dangerous, for humans to do, but there is no shortcut to the kind of intimacy Park aspired to with his subjects. Rigorous self-abnegation is the only way to see tigers in their natural state: hunting, feeding, relaxing, courting, playing with their cubs. "When I [am] humble," Park told one interviewer, "I can see nature more deeply."

Park's humility is evident throughout these pages, and it is often attended by a striking lyricism that, even in translation, reminded me that most of what has been written about Siberian tigers comes to us through English and Russian. I had the feeling, reading *Great Soul of Siberia*, that it could not have been written by a Westerner. Unlike most tiger researchers, Park is local: raised in the Sobaek Mountains of South Korea—former tiger country—he is the product of a culture whose poetic tradition dates back at least two thousand years. Perhaps it is this, combined with his background in literature, that explains a lyricism, tone, and intimacy with the landscape and its inhabitants, human and animal alike, that I haven't encountered in other tiger literature. Park's distinctive voice, coupled with the fierce and monastic dedication of his vigil, brings us so close to this extraordinary animal that we can feel, as he did, the tiger's breath on our face, her whiskers brushing against our hand.

It may seem strange to use the word "personal" when talking about tigers, but this is the overriding feeling I had while reading Park's book. Spending time in his wild, frozen world, I came to realize that "personal" is another way of saying "intimate," which in this context also means "mammalian." Park is writing about tigers, but he is also writing about families—the domestic life of tigers in the wild. And it is through these relationships that we gain a window into the heart and soul of Park himself, a family man who aspires also to be a "Man of Nature." Viewing these animals through Park's inclusive, empathic lens, we are privy to scenes that are both familial and familiar: birth and death, satisfaction and desire, tenderness and cruelty, teaching and learning, mischief and discipline, comedy and tragedy—the full spectrum of family life as it pertains not only to *Homo sapiens* but to all higher mammals. As a vicarious witness to these private scenes, many of which have never been observed, much less filmed, in the wild before (a father keeping watch while his cubs feed; a mother filling in as a sparring partner for a cub who's lost a sibling), I felt the boundaries between "us" and "them" begin to waver and dissolve. I found myself wondering how similarly we might behave if we lived outdoors all year round in a forest where temperatures drop below –30°C and we had to hunt large game to survive; how we would cope trying to feed a family while being hunted ourselves.

And it is this, the generous invitation to see and feel beyond our own boundaries and to explore the harsh and tenuous reality of one of our shared planet's most compelling creatures, that is the greatest gift a Man of Nature like Sooyong Park can give us. *Great Soul of Siberia* offers a unique portal through which we can find common ground—and perhaps common cause—with these rare and elusive animals that seem so profoundly different from us, but with whom, as Park so eloquently demonstrates, we share so much.

PROLOGUE

SINCE 1995, I'VE been researching and observing wild tigers in Manchuria, the North Korean border region, and Ussuri (officially known as Primorsky Krai), which together form the southeastern corner of Siberia. I've spent half of each year wandering forests and climbing mountains in search of tracks and the other half in underground bunkers filming tigers in -30°C weather. Twenty years of waiting for tigers that seldom come has been a frustrating and humbling experience, but ultimately a rewarding one that has yielded seven documentaries on Siberian tigers, including *Siberian Tiger Quest*. When I first began, there was less than an hour's worth of wild Siberian tiger footage in the world, but twenty years later, I've amassed close to one thousand hours of footage. The observations I've made with my own eyes are three or four times the length of my recordings.

Blending into nature and staking out for long periods are no small undertakings. When I'm confined to an underground bunker in the wild, I have to take care of everything in that space: go to the bathroom, melt frozen rice balls for meals, and brave the cold Siberian winds. After six months of not being able to shower, shout, or turn on the light, I begin to identify with prisoners in solitary confinement.

When a tiger appears after an interminable wait, it's as if I'm seeing the face of a lover I've been dreaming of. My heart races as I'm overwhelmed with joy, and time stretches on like eternity in those few brief seconds. Then, suddenly, joy is replaced by terror. The tiger notices something is out of place and heads straight toward me, its eyes burning blue in the dark. Footsteps approach, snow crunching under the tiger's paws. The sound stops. I feel the tiger's hot breath on my face. The fragile pendulum of existence swings between life and death before my eyes.

But in the end, it isn't the fear of my physical reality in the wild—the cold, the beasts—but an absolute sense of aloneness that freezes me to the bone in the bunker. As I tremble in the two-square-meter space like someone having an attack of claustrophobia, I realize that, unlike tigers, humans are a species meant to live alongside others.

There are five species of tigers on our planet: the Siberian tiger, the Bengal tiger, the Indochinese tiger, the Sumatran tiger, and the South China tiger. All except the Siberian tiger live in tropical regions. The Siberian tiger lives in Ussuri, Manchuria, and the Korean Peninsula. Russians call it the Amur tiger, and the Chinese call it the Northeast tiger (as opposed to the South China tiger). They're Manchurian tigers when spotted in Manchuria, Ussuri tigers when spotted in Ussuri, and Korean tigers when spotted in the Korean Peninsula, but they all belong to the same subspecies known as *Panthera tigris altaica*.

Siberian tigers are very reclusive compared to their tropical counterparts; they avoid contact with humans at all costs and live secretive lives in vast mountain ranges. Chancing upon a tiger in the woods is like finding a needle in a haystack. The luckiest tiger researchers get to see wild tigers maybe once or twice in their entire careers. It's no surprise, therefore, that a great part of the Siberian tiger's behavior remains a mystery. Next to nothing is known about how tiger families are formed and dissolved, such as the behavior of females with cubs or the relationships between males and cubs. But when we do happen to see Siberian tigers in the wild, their talent for hiding, their wisdom to gauge the situation and know when to back off, and their bravery to fight to the end when they must are wondrous to observe.

Ussuri is also inhabited by indigenous peoples who worship the Siberian tiger. They believe that all life forms of the forest, water, and earth have spirits. Their god of the forest is Amba, or "the strongest one," whose physical form is the Siberian tiger.

The forests of Ussuri, Manchuria, and the Korean Peninsula have long been the backdrop for both humans' and tigers' difficult fight for survival. This wilderness is becoming increasingly inhospitable due to logging, development, extreme weather, and poaching. Rifles have replaced spears, booby traps have replaced rifles, and landmines are emerging as the next poaching weapon.

Once referred to as the "Amazon of Northeast Asia," the Ussuri forest is now poachers' heaven. Dozens of the 350 remaining Siberian tigers disappear from the planet each year. This species, considered in the past to be the bravest and most sacred on earth, is walking down an irreversible path to extinction at the hands of mankind. The once-thriving Siberian tiger population of 10,000 is now reduced to 350. The population of indigenous Ussuri has also been reduced from several hundred thousand to a mere ten thousand who continue the lonely fight to keep their culture alive.

This book is a record of three generations of tigers—a female tiger named Bloody Mary, her cubs, and her cubs' cubs. It is also the story of the native Ussuri who worship Siberian tigers. I hope this book does what little it can to shed light on Siberian tigers and the struggles of the native Ussuri and give them all a better chance at survival.

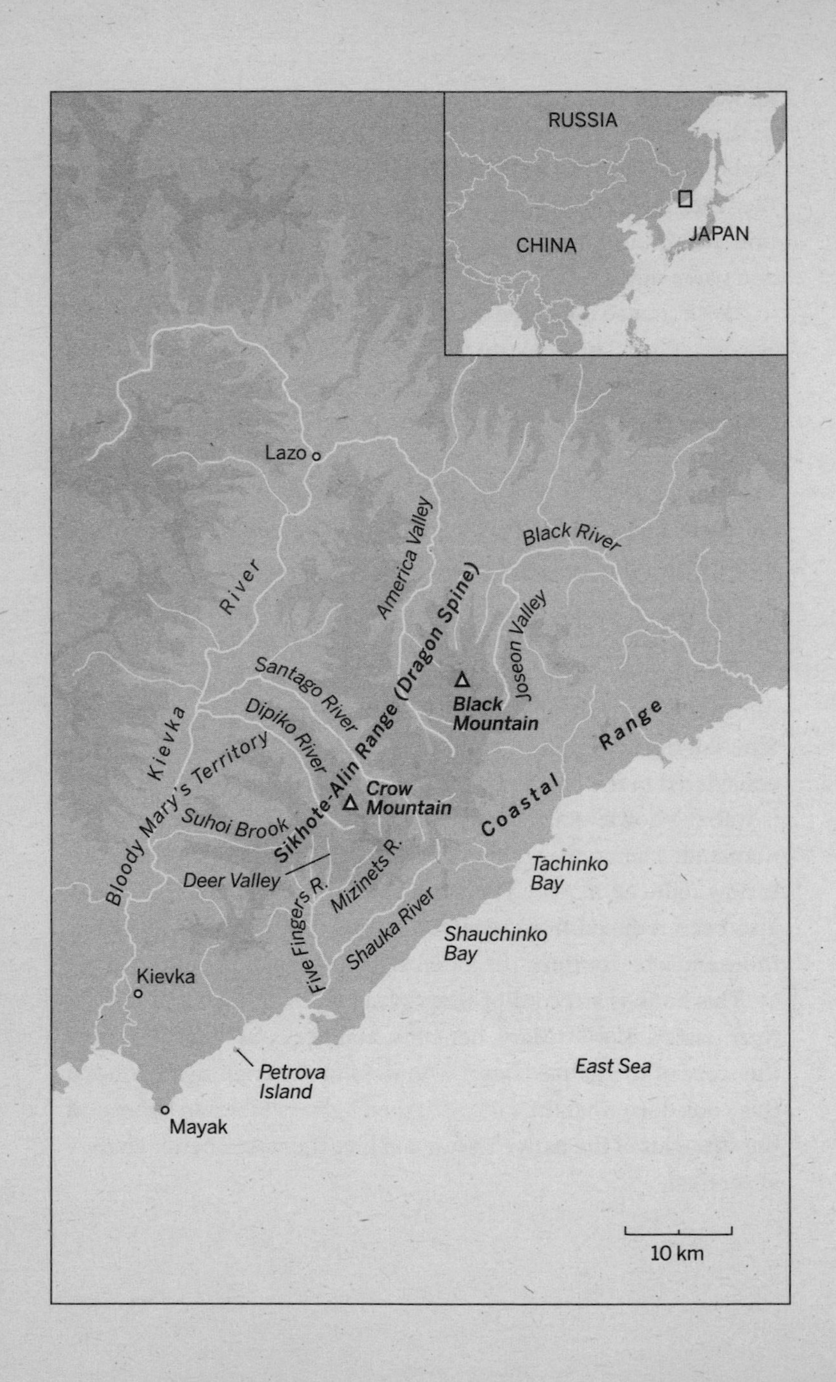
RUSSIA
CHINA
JAPAN
Lazo
America Valley
Black River
River
Sikhote-Alin Range (Dragon Spine)
Joseon Valley
Black Mountain
Santago River
Dipiko River
Kievka
Bloody Mary's Territory
Crow Mountain
Coastal Range
Suhoi Brook
Deer Valley
Five Fingers R.
Mizinets R.
Shauka River
Tachinko Bay
Shauchinko Bay
Kievka
Petrova Island
Mayak
East Sea
10 km

PART I

Bloody Mary

CHAPTER 1

Tigress in the Basin of Skeletons

SOARING, MAJESTIC SWELLS of dark blue raged toward the shore. The breeze was cool. Winter was finally coming to an end. Nearly all of the snow had gone, except on the mountain peaks. The forest and hills were cloaked in gray and mustard yellow, the dreariest palette of the year. Dark and light swatches stretched on like a rag patched and mended too many times, and the ice in the brooks and rivers cracked and broke apart as the muddy currents carried it away. The arctic warbler had not begun to sing yet, and there were no flourishes of green, but the tremendous energy of life heaved just beneath the surface of the faded earth as it pushed through its final toil.

The perilous Tachinko Cliff towered over the sea that threatened to pull the onlooker into its dizzying waves. An eagle circled the sky and landed on a peak on the other side of the cliff. Heat shimmers rose in the warm sunlight amid azalea buds ripe and ready to burst.

Click, clack. Click click clack.

A couple of stones rolled down the cliff on the other side. The eagles were alight. Two Ussuri deer carefully made their way down the steep cliff. In Korea, they're called plum flower deer for the white spots that blossom on their red coats in the early summer, but at this time of the year, their coats are as shabby and yellow as the lifeless skin of the Ussuri forest. The two deer looked scraggly and weak. The largest of the sika deer species, Ussuri deer once also lived throughout Korea and northeastern China, but their population has diminished over the years and they are now found mainly in Ussuri in the southern panhandle of Russia more commonly known as Primorsky Krai.

Every year, the Ussuri deer that have made it through the winter come down to the shore in the early spring. Before the first new leaves, when food is still scarce, the deer survive the last push by filling their bellies with seaweed or kelp. This also replenishes their salt intake, which is typically low during the winter. Their trips to the shore begin when the snow starts to melt and peak around May when the deer give birth to their young, as pregnant deer need more sodium to keep their growing fetuses strong.

More stones rolled down the cliff, then a heap of rocks crashed to the bottom. One deer fell off the cliff in the mess of rocks. It hit the cliff once on its way down and fell into the ocean. When the deer surfaced again, it could hardly move. With only its head above water, the deer was caught in the waves. It was a young stag that had just begun to grow antlers.

When I approached the stag, it was struggling to climb onto a rock. It used every drop of strength to push itself up, only to be swept under again. Stefanovich, the forest ranger I was traveling with, and I pulled the deer onto the rock, where it sprawled and gasped for air. It looked at us with pleading eyes.

Eagles were perched on every peak along the cliff. Like locals well versed in the ways of these parts, they lay in wait, looking off into the distance with an air of indifference.

The deer was still wide-eyed and gasping for air, but there was no hope for its survival. Stefanovich raised his gun to put it out of its misery, but I stayed his hand. I didn't like the sound of gunfire. Perhaps this came with the territory of moving soundlessly in search of tigers—I didn't want to disturb the quiet flow of nature. The deer was in pain, but it wasn't our place to intervene. In the end, the birds perched on the peaks would get to him.

ONE BRANCH OF the Sikhote-Alin Mountains stretches south along the eastern coast of Ussuri through the Lazovsky District. Tachinko Shore lies in the middle of this coastal range. To the left and right of the shore are steep cliffs where wild goats live; over the ages, the powerful waves of the East Sea have dug out a crescent beach. About one kilometer down this beach of coarse sand and pebbles is a vast, thick valley of cork oaks.

We found deer tracks on the beach covered with tiger pugmarks. It looked like the tiger had pounced on the deer. Their tracks led to a spot where the sand had been churned up during the struggle. The tiger had hidden on the edge of the cork oak forest and caught the deer in just three tries. While the tiger pounced three times over a mere fifteen meters, the deer had managed to take only one leap. This was unusual. Such swift hunting is normally only possible when the deer is lying down to rest very close to the spot where the tiger hides in wait.

But there was no evidence of the deer lying down. It had been headed for the beach, but even then, could take only one leap for its life. It had been a perfect stakeout and a clean attack. The tiger had demonstrated cunning camouflage and agility rather than brute force.

Male Ussuri tigers use tremendous power when they hunt deer, boars, bears, and other large animals. But the relatively small and sensitive female tigers use a more effective weapon—trickery.

The sand looked rusty. The deer had bled profusely. The blood in the sand had not yet congealed. The deer must have died the previous night or early that morning. The tiger had dragged it into the cork oak forest. After killing their prey, tigers typically take it to a secluded area where there is water to drink. The tiger had eaten its deer by a small brook in the forest. There was a great deal of blood by the brook as well. It dawned on me that this was the handiwork of Bloody Mary. The locals named the tiger Bloody Mary for her habit of soiling the earth with blood whenever she took down a deer or boar.

Bloody Mary killed by delivering a fatal bite to the neck like any other tiger, but she always added an insurance bite. She would clamp down on the neck of an already dead deer and give it a hard shake, boring large holes into its throat with her long conical fangs. The arteries would tear and blood would spill. She had a cautious, tenacious personality and liked to be sure she'd done the job right.

Nobody in the area had ever seen this tiger, but it was assumed by the bloody remnants of her hunts that she was ruthless and cruel. That's how she'd earned the same nickname given to Queen Mary I, the British monarch who executed countless Protestants in the sixteenth century. But Bloody Mary the tiger was merely vigilant, not ruthless, when it came to humans. I had never heard of anyone being harmed by her. Rather, thanks to her persistence and her diligence at keeping her distance from humans, she had been able to stay safe and raise her young well. She excelled at detecting dangerous contraptions laid by man and was a skilled hunter more than capable of feeding her young, even when she had a large litter.

She was as keenly wary of humans as she was deeply attached to her territory. So elusive was her presence that nobody had seen so much as her shadow, but when she sensed someone in her territory, she furtively circled the area and kept an eye on them until they left. While we were there, she had been patrolling her territory on the Tachinko Shore in southeastern Lazovsky. She had likely watched us observing the Ussuri deer and the wild goats. She had never once revealed herself to us, but often left traces nearby.

Like a suspicious mother bird who senses a threat to the eggs she is brooding and circles the nest to investigate, she was lurking around our expedition. Maybe we had crossed paths in the forest. I couldn't see her, but she could see me. She would have learned to tell me apart from other humans and knew that I was not a hunter. She was, after all, an Ussuri tiger. And Bloody Mary at that. If she had been any other tiger, I would not have had the nerve to wander the forest.

A patch by the brook was littered with tufts of deer fur that Bloody Mary had ripped out with her front teeth. It was just like a female tiger to pull out the fur so meticulously before eating. The deer's head was cast aside pathetically, its nose stuck in some leaves and its filmy dead eyes staring at nothing. The clouds in the sky and shadows of trees faded in its pupils. I looked closely and saw that its molars were yellow and ground down. It was an old stag. Its entrails strewn about the ground did not contain much. It seemed the stag had fed mostly on dried oak leaves, but not enough of those either. Its face looked haggard. Even though it was a scrawny deer with not much meat on its bones, Bloody Mary had left bits of it uneaten. She hadn't even touched its head, and there was meat left here and there on its ribs and legs.

The brook had thawed almost completely, and clean water babbled along. The parts of the ice that were still frozen had air bubbles growing under spots of sunlight; other parts were full of round holes. Bloody Mary had drunk from the brook after feeding on the deer. Prints of her two front paws were stamped neatly and cleanly in the damp earth by the brook. Tigers' size, age, and sex can be determined by their pugmarks and the lengths of their strides. The width of the front paw pads is an especially important indicator. If the width is greater than 10 centimeters, the tiger is most likely male. Few females have pads that wide. The paw prints by the brook were 9.7 centimeters, large for a female tiger, but the depth of the prints suggested the tiger was light. It was a smart, sensible female tiger, small but good at raising her young.

After a drink of cool water, Bloody Mary had crossed the brook and gone into the cork oak forest, a snug basin with Tachinko Shore to the southeast. Deer have to pass through this forest to reach the shore. Some get to the shore by climbing down the cliffs to the left and right, but those paths are too steep and not popular among the deer.

The large basin is densely populated by cork oaks. The yellow-brown leaves on the ground, the dark bark of the cork oaks, the dry blades of grass, and the field of reeds on the edge of the basin provide perfect camouflage for tigers stalking and hunting prey. Some of the deer that feed on the dry oak leaves in the basin on their way down to the shore meet their ends here. The ghostly white skulls and bones of deer that tigers have preyed on over the ages are scattered throughout the forest. On a sunny day with a gentle breeze, the forest is inviting. The dry leaves cover the ground like a cozy blanket and create a languid, sleepy atmosphere. It is a soporific place. But on days when the fog rolls in from the shore, the dank bark of the wet cork oak and deer bones look eerie in the milky darkness.

Sometimes it seemed the bitter ghosts of the dead deer were rising from the ground and gazing out through the thick fog, and other times I expected Bloody Mary to approach soundlessly from behind and pounce on me at any second. In those moments, the forest was transformed into a ghostly, wet swamp I did not want to cross. It is probably worse for the deer that have to pass through the basin every year—hence its name: the Basin of Skeletons. I found at least two skulls from deer that had died within the previous month or so. I imagined they were Bloody Mary's kills. Only then did I know why she didn't have to lick the meat clean off the deer bones. Spring was a terrifying time in Tachinko for the deer.

The spring that felt slow to come finally arrived, but the fallen deer did not live to see it. The southeastern region of Lazovsky, including the Basin of Skeletons, was a harrowing place for the deer, but perfect for Bloody Mary to raise her cubs, especially in the spring. Her choice territory was another reason she was known for being a good mother. I started to wonder what she looked like.

I didn't know then where this journey would take me. Never could I have imagined that I would one day feel the warmth of her breath and her long, stiff whiskers on the back of my hand, and that I would be there, as well, to witness her death.

CHAPTER 2

The Traveling Base Camp

AMUR ADONISES ARE the first flowers to bloom when spring comes to Ussuri. A cluster pops up through the melting snow in sunny spots. Wildflowers follow the Amur Adonises, carefully pushing up their buds, and little spring blossoms of red, yellow, and purple grow in small colonies like colorful landmasses on a world map. The forest turns into an endless garden of wildflowers. Around this time, squirrels and chipmunks emerge from their winter naps. Badgers wander about looking for wild plants to dig up so they can feast on the bulb roots. Asian black bears come out for a walk with their cubs, who were born over the winter. Mother bears lift rocks to find ants and caterpillars for their cubs to eat, and the cubs wrestle in the velvety green fields and break their winter fast by feasting on new leaves. Wild boars roam in groups, digging the fertile soil for fat slugs and worms. The boar herds can be counted on to turn the wildflowers into a chaotic mess, as if a dinosaur had rampaged through.

My long winter stakeout had finally ended, and I was to shift gears and start tracking tigers to map their territories and routes. From May through September, I tracked the tigers' droppings in the forest and compiled a map of their migration routes to decide the best locations for my next stakeout. Then I spent October through April of the following year in my hideout waiting for the tigers. Half the year was spent on field expeditions and the other half on stakeouts, two ways of studying nature.

FIELD EXPEDITIONS ARE essential to the success of a stakeout. Through field expeditions, we pick up the various details that will determine where, when, and how in the vast, deep wild we should conduct our stakeouts. By following the traces of an elusive animal in our summer field research, we prepare for winter field research: staying in one spot until our subjects reveal themselves.

The area where we observe wild tigers is Lazovsky, southeast of Ussuri, home to the Lazovsky Nature Reserve. Nature reserves in Russia are called *zapovedniks*, and they have the strictest regulations of all environmental conservation areas in the country. No one is allowed in zapovedniks without permission, and all plants, animals, and minerals in the reserves are protected rigorously. So when I think of the word "zapovednik," the first thing that comes to mind is "authorized personnel only." Of course, even authorized personnel must receive clearance each time to get in. It would be impossible to fence in the vast area, but trespassing is just as prohibited as on privately owned land.

Rangers, the law enforcement authorities of the woods, can arrest trespassers and open fire if they flee and can interrogate and indict perpetrators. It's illegal to enter a zapovednik, and those who poach, fish, harvest, or log in the reserves are subject to heavy punishment. Because of these strict regulations, wild tigers that have almost completely vanished from Korea and Manchuria have been able to maintain a minimum viable population (the minimum population required for a species to survive without compromising genetic or ecological diversity and health), albeit precariously, in Ussuri.

There are 101 nature reserves in Russia, making up 1.6 percent of all Russian territory. Three of these nature reserves located in Ussuri near the Tumen River are home to wild tigers. The smallest is the Ussuri Nature Reserve, the second-largest the Lazovsky Nature Reserve, and the largest the Sikhote-Alin Nature Reserve. Zov Tigra National Park (812 square kilometers), although not a zapovednik, was recently established to protect wild tigers.

The Ussuri Nature Reserve is 578 square kilometers. Only four or five tigers live there because the area is not very large and it is cut off from the Sikhote-Alin Range. Most of these tigers have not made stable homes of the nature reserve, but lead nomadic lives on the outskirts. In reality, only Sikhote-Alin and Lazovsky are real homes for tigers.

The Sikhote-Alin Nature Reserve is located in the middle of the Sikhote-Alin Mountain range that runs along the eastern coast of Ussuri. It was the first nature reserve established to protect tigers, and at 4,014 square kilometers, the largest. It's home to between twenty and twenty-five wild tigers.

Lazovsky Nature Reserve, on the other hand, is located south of the Sikhote-Alin Mountains. The nature reserve itself is 1,210 square kilometers, but the entire expanse of the Lazovsky District, including the outskirts where tigers also live, is about 4,700 square kilometers. Between eight to twelve wild tigers live in the Lazovsky Nature Reserve and its outlying areas. Bloody Mary was one of them.

WHETHER WE'RE ON a field expedition or stakeout, we always begin at the base camp. Base camps are generally set up in a village or a mountain lodge near the research area. When we are conducting research in many different areas, we set up shop somewhere in the middle.

On this excursion we were studying Bloody Mary, whose territory spanned from the southeastern coast of Lazovsky to farther inland, so our base camp was located in the southernmost Lazovsky city of Kievka, seventy kilometers away from downtown Lazo. Since before recorded history, this part of Ussuri has been populated by the Udege and Nanai peoples. They are the descendants of northern nomadic tribes that shared linguistic and ethnic roots with the Manchu. The Udege were hunters who dwelled mainly in the forest and opted for a nomadic life instead of building villages. They moved in clans and followed game all over Ussuri. The Nanai, on the other hand, built riverine villages, hunting in the winter and fishing in the summer, and were nicknamed "fish skin people" because they made fish skin clothes and shoes. Part of the Ussuri ecosystem, they have hunted and fished for generations here.

This village of about a hundred households was once occupied by native Ussuri peoples, but was taken over by Ukrainians from Kiev in the nineteenth century, who changed its name to Kievka. The native Ussuri population has been nearly obliterated. Only a few families live in the village of Mayak about fifteen kilometers outside Kievka. Most of the villagers there make a living as medicinal root gatherers.

Our base camp at Kievka was a small wood cabin. It had a spacious entryway, one large room, one small room, and a kitchen through the entryway door. A wood stove was built into the wall where the three rooms met, distributing heat evenly among them. The cabin received a lot of light in the morning through its many windows. It was our home and command center. Here, we collected data, fixed our equipment, and put together supplies to distribute to our stakeout posts. We recharged ourselves as well as our equipment batteries at the base camp. When I returned to the base camp from a long stakeout or camping in the woods, this cabin felt like home.

We also had an expedition base camp. Ussuri tigers lead very itinerant lives. One or two hundred kilometers is nothing to them.

Since they cover such vast distances, we had to have a base camp we could move around. So we bought a secondhand four-wheel-drive truck from the Russian military and named it "the Ural." It guzzled gasoline but was as sturdy as the Ural Mountains.

We pulled the truck bed apart and installed a four-meter steel container called a *kun*. Stefanovich and Valosia, the two forest rangers who assisted me with our expedition, were thrilled with the project. It's very exciting for people like us, who live in the woods, to have a traveling cabin. We installed a four-person bunk bed, a wood stove, and lighting fixtures in the kun. Once we added a chimney on the roof for the wood stove and cut out windows on either side of the container, we had ourselves a respectable traveling base camp. Stefanovich was assigned the task of driving and maintaining the Ural.

Once we were out in the woods in the Ural, the survey usually went on for about a month. We searched the forest for tiger pugmarks by day and returned to the Ural to sleep at night. When we camped for longer periods, the Ural dropped us off on one side of the mountain and waited for us at a designated spot on the other side. The expedition traveled over the mountain on foot, researching during the day and pitching camp at night. We rejoined the Ural when our journey over the mountain was complete.

Collecting data in the field, we were often caught in thunderstorms or extremely cold weather. On such days, it was a comfort to have the Ural with us. When we crawled into the kun and fired up the stove, it became as warm as a hearth inside. We hung our wet clothes to dry, made food, drank vodka, and rested up for the night. The cramped space inside the kun seemed so cozy on such occasions. After a month of field research, we returned to our base camp in the village. We recharged for a week or two and then set out in the Ural again for another part of the forest.

The Ural was a Russian military truck, but even it couldn't avoid breaking down when we drove it through the rugged mountain terrain. There are no service stations in the mountains, so if you're not confident you can fix a broken car, you don't dare make that drive. But like all Russian drivers, Stefanovich and Valosia were not only great drivers but also expert mechanics. Even when their cars broke down in the village, they did the repairs themselves. They would find a car that was too worn for use, park it on one side of the yard, and take parts from it to fix their cars. Nothing went to waste. They hoarded even the smallest bits to reuse later. Frankly, they were too poor to even afford new car parts. This made them resourceful mechanics.

One time, the Ural's engine spurted a puff of white smoke and stopped on the way up a mountain path. We opened the hood to find the fan belt had snapped and half the coolant had evaporated. I was worried because we didn't have a spare fan belt, but I had faith that Stefanovich and Valosia would think of something. And I was right. Stefanovich took a look around the area as if it were no big deal and asked me for my camera. I surrendered my camera to him, not knowing what he was up to. He undid the camera strap and fashioned it into a fan belt with some fishing wire. He had a terrific fan belt in no time. He fetched some clean water from a nearby stream to refill the radiator and installed the new fan belt. The Ural let out a long blast as it drove on, as tough as ever. Thanks to their excellent driving and knowledge of car mechanics, the Ural carried out its mission no matter how far we traveled in pursuit of the Ussuri tigers.

CHAPTER 3

The Spirit of the Sikhote-Alin

WE SET OUT from the base camp in the Ural. About fifteen kilometers north along the new road to downtown Lazo, we came upon a small forest path leading to the Sukhoi Brook. Valosia and I picked up our backpacks and hopped off the Ural, which turned around and drove back the way we had come, leaving a cloud of dust on the unpaved road in its wake.

We had found traces of Bloody Mary and her cubs a few times in the past, but no one had ever seen her. One would think the medicinal root gatherers who went into the forest every day might have run into her, but her characteristic cautiousness had kept her hidden. We asked people living in the villages near Bloody Mary's territory if they had spotted pugmarks of tiger cubs, and discovered that Bloody Mary had had two litters. She'd given birth to two cubs seven years earlier that she'd safely brought up to adulthood, followed by a large litter of four cubs four years earlier. She'd lost one cub in the process, but there were signs the remaining three cubs had stayed close to her until they left her care.

A female Ussuri tiger first has cubs when she is three years old. The cubs are raised for two to two and a half years before they leave their mother's care. Based on this, we conjectured that Bloody Mary was a seasoned tiger of around ten years old. According to our calculations, she was out there raising her third litter, assuming she'd mated. We were curious as to whether she had had cubs, and if so, how many. The focus of our research that year was Bloody Mary. We imagined it would be difficult to see her in the flesh, but we planned to follow her traces, study them, and get as much information on her as possible.

We intended to spend the month of May collecting data along Sukhoi Brook and into the upper reaches of Mizinets River. We would recharge for two weeks after that and then resume our expedition along Santago Valley to Shauchinko, then down the coastal mountains from Tachinko to Mayak Village in August. We would spend September building stakeout trenches based on the data we'd collected, then head into our long six-month stakeout from October to March the following year.

The leaves on the trees were starting to unfurl. Myriad shades of green gave the forest new dimensions. Nut pine needles were the deep blue of the ocean, while the larch colony was the color of a light blue-green coral reef. The new leaves on the maple trees were yellow-green, and the oak colony was light green. The willow and poplar leaves were icy blue with a hint of milky white. The white base of the birch had shed its protective winter layer. I peeled off strips of bark so white I thought my fingers would turn white from touching it. The new bark, infused with blue, grew beneath the old.

The Udege and Nanai share an inextricable bond with birches and willows. In the spring, traditionally, they scraped off the inner bark of birches and mixed it with grain to steam and eat, and picked the chaga mushrooms that grow on the sap of birches in the autumn to extract oil, make wine, and boil tea. They chopped down birch trees to build their houses and make firewood, and peeled off the dried bark to use as tinder or make household paraphernalia such as baskets and cribs. Willows were used to make dog sleds and fishing poles, and the bark was dried to make ropes. They brewed the willow bark, now the well-known raw material of aspirin, and drank the tea to soothe their aches.

The Udege and Nanai still believe that the spirit of a person comes from trees and when one's time in this world is up, men's spirits return to willows and women's to birches. They believe that the trees go through a period of rest and are reborn as new spirits. And so they see this world as a place where spirits pass through eternal cycles, where there is no death or sadness. To the Udege and Nanai, everything in the world is a living thing that gives and receives energy, and that energy is temporarily borrowed from nature and must be returned when the time comes—life and death are repetitions of these cycles and a circulation of energy.

The indigenous people of Ussuri who lived by the river must have felt a close bond with willows, and the vast expanse of birches would have been a friendly sight to those who lived in the forest. They were born of trees, lived among them, and returned to them. So they considered the forest part of their ancestry and history and referred to themselves as Udege, "the people of the forest," and Nanai, "the people of the earth."

We climbed a small ridge and saw the Sikhote-Alin Range in the distance. Far north along the green ridge was what the Ussuri call Black Mountain (officially known as Mount Chyornaya). Black Mountain is rockier toward the top, so little grows around the summit except for a few coniferous trees and shrubbery. The dense forest in the lower altitudes, on the other hand, is vibrant with life, and the Ussuri call it the Bright Forest.

Two hundred kilometers to the north and south of Black Mountain lie treacherous mountain terrains. The Udege refer to these as the Dragon Spine. Looking down from the peak, one can see the imposing boulders that form the ridge of Black Mountain. The ridge resembles an enormous dragon that has died and left its strong, coarse spine after its flesh and skin have turned to dust and blown away. The plunging, beautiful valleys of Koropad lie to the northeast of the Dragon Spine with the Bright Forest at their feet; to the southwest is the top of America Valley, which flows down from Black Mountain. Beyond that, Santago Valley (Udege for "third largest valley," which it is in Lazovsky, after America Valley and Koropad) unfolds like a dream. Santago Valley and the area to its south formed Bloody Mary's territory.

The Lazovsky District maintains a carrying capacity of about ten wild tigers. They travel around the nature reserve and its outlying areas, claiming the 4,700-square-kilometer expanse as their home. The tigers' territories are divided along Black Mountain in central Lazovsky. During our expedition, it was difficult to tell exactly where the boundaries were, but the area was roughly divided into four quarters, each occupied by one female tiger. The area west of Black Mountain belonged to a tiger in her prime, while the northwest held an elderly tiger, the northeast a Koropad tiger, and the south Bloody Mary. Each tiger had a territory of about 500 square kilometers. The territory of male tigers is about four times that of female tigers.

Bloody Mary's territory, which spanned from the southeastern coast of Lazovsky to beyond the Dragon Spine, was over 500 square kilometers. Collecting data in Bloody Mary's territory took a long time because the research perimeter was vast and the terrain forbidding. Also, the feeding patterns and migration of the ungulates—deer, boars, and other hooved animals that tigers like to prey on—varied each year, which made it difficult to pinpoint Bloody Mary's favorite routes and haunts.

But there were certain areas that she was sure to visit often—the territorial borders. Tigers are very competitive and protective when it comes to territory and do not take kindly to trespassers.

A female tiger's territory doubles as her home, where she gives birth to and raises her cubs, and her hunting grounds, where she must find enough food for her young. If female tigers happen across a trespassing tiger, they get into a fierce brawl. But most tigers make sure they never run into each other by taking precautions like going on regular border patrols and maintaining territorial markings that serve as warning signs. They dig holes and leave droppings that contain their own unique chemical scent markers or mark large trees by engraving long vertical claw marks on the trunks. They also rub their necks against the base of tree trunks to leave clumps of their fur or spray the trees with pungent scents.

These territorial markers are how tigers make their presence known. For male tigers and their families, it's a means of communication. For rivals of the same sex, these are warning signs. Around Black Mountain where the territories met, the four female tigers were always waging fierce psychological warfare.

There were four or five other female tigers in this region, but they were young—independent but inexperienced. They wandered from area to area with no established territory of their own or had been pushed out to the edges of the nature reserve. Luckily for the young female tigers, most of them were the offspring of the four females who had settled Lazovsky, so they could skulk around the edges of their mothers' territories unharmed. Young male tigers, on the other hand, faced a greater challenge. They had to avoid the Great King, the largest and strongest male in Lazovsky. Young male tigers ended up wandering all over Lazovsky or were chased to the outskirts while the Great King mated with the four female tigers and reigned over all of their territories.

Since ancient times, the native Ussuri have referred to the strongest male in its prime as the Great King. Siberian tigers' stripes are faint and thin when they are cubs, but grow thicker and more distinct as they reach adulthood. Once they are fully grown, a vivid pattern that resembles the Chinese character 王 (king) appears on their foreheads and 大 (great) on the scruffs of their necks. Unlike tigers in tropical regions that have stripes as thin and sharp as if they were painted on with blades, Siberian tigers have uniquely thick, simple stripes. Such patterns are pronounced on male tigers that are significantly larger than female tigers, especially on the largest and strongest male.

THE UDEGE AND Nanai are animists; that is, they believe that everything on earth possesses a spirit. They are the last inheritors of the religion and cosmology of the Tungusic people, an ethnic group that inhabits eastern Siberia and Manchuria and that includes the Manchu ethnic minority in China. They believe the world is divided into the upperworld, middleworld, and underworld. The upperworld is the realm of the sky where the great spirits live, the middleworld is the realm of the earth occupied by humans and spirits of nature, and the underworld is the realm below where the spirits of the dead dwell. The three worlds are connected by the Engzekit River, which is rather like the River Styx in Greek mythology.

The god of gods, Enduri, and the other great spirits live in the upperworld. Enduri governs the balance between the realm of the sky and the realm of the earth where humans live. To maintain this balance, Enduri battles evil such as cold and lightning and sends the great spirits of water and forest to the middleworld as his messengers.

Temu is the great spirit of water. The sea and rivers are his responsibility. To carry out his duties, Temu also sends messengers. One of those is the kaluga, a sturgeon that swims up the Amur River from the Sea of Okhotsk each May. Kalugas are huge, over five meters long and weighing easily over five hundred kilograms. The Nanai who live by the Amur River fish the kaluga, but out of respect, they do not kill more kalugas than necessary. They are especially careful not to overfish female kalugas carrying eggs, and they perform a ritual offering to atone and appease the spirit of a dead kaluga.

The middleworld is the realm of the earth, where nature and humans live. Nature is a being that possesses mystical powers, and all things in nature have their own spirits. The spirits, such as those of water, fire, wind, rock, and fish, take the form of animals. This animist philosophy is deeply entrenched in the Udege and Nanai way of life.

Amba is the god of the forest, whom Enduri sends to the middleworld to preserve the balance between humans and nature and to keep the cycle of energy flowing. Amba communicates with all the spirits of the forest and shares energy with humans. The Udege and Nanai worship Amba as the god of the forest. Udege for "the strongest one," Amba refers to the tiger who rules the Ussuri forest. The strongest of the strong is the Great King.

The Sikhote-Alin, where the Great Kings live, has been home and spiritual stronghold for generations of native Ussuri. They call it Kunka Kyamani, or "sleeping spirit." One day, an Udege from Lazovsky was journeying through Kunka Kyamani when he came across an enormous male tiger. Since then, the Udege have called this tiger Kunka Kyamani Khajain, or just Khajain for short. Khajain is the current Great King of Lazovsky—the spirit of Sikhote-Alin.

The Great King usually mates with four or five females. On average, he maintains an area of around 2,000 square kilometers, and the circumference of his territory is about 200 kilometers. But no one knows just how big his territory really is. We cannot know if Khajain has claim over all the 4,700 square kilometers of Lazovsky as 'Tail', the Great King before him, did, or if he also roams farther north to the Chuguyevka region, as Kuchi Mapa,

Tail's predecessor, did. Kuchi Mapa traveled 400 kilometers from southern Lazovsky to northern Chuguyevka, and a Great King in the Khabarovsk region was once spotted at Lake Baikal, 2,000 kilometers away. Another Great King was seen in Yakut, in central Siberia, 2,500 kilometers away from his territory. I call these far-traveling tigers "Gwanggaeto tigers," after Gwanggaeto the Great, a Korean monarch from the fourth century famous for his territorial expansion. One can never know for sure how far the territories of Gwanggaeto tigers extend. Perhaps their boundless ambition renders them fearless and adventurous. It is a characteristic one only sees in the Great Kings.

CHAPTER 4

Fishing Tiger

THE SMELL OF pine nuts became more intense the farther we climbed along the ridge. Before long, a forest of nut pines came into view. Every new branch was adorned with new buds that made the forest vibrant with color. The red, musty pine nut pollen, a deeper hue of red than pine pollen, flew at me in the wind. The pollen fertilized the purple pine flowers, which bore little pine nuts that ripened in the autumn over a year later. This meant the golf ball–sized pinecones had been fertilized the previous year. They would ripen in the fall and become a food source for ungulates in the winter.

I counted three or four pinecones on each branch. Some branches had five or six. The nut pines had had a good year. It was important that I note the yield on the nut pine trees, oaks, and wild walnut trees as I conducted my field research. Ungulates have no trouble roaming the woods in search of food between spring and fall, but must resort to scavenging these areas in the winter for pine nuts, acorns, and walnuts buried under the snow. Tigers follow the tracks of the ungulates to these areas. The food yield in different regions is a good basis for choosing hideout spots.

It doesn't matter, however, how much fruit the nut pines or oaks bear if they're near villages. In the fall, the poor village people come to pick pine nuts and gather acorns. So it's important to find places, such as Crow Mountain, that have plentiful food but are difficult for people to reach. In the fall, we stay in places like this for a long time to research them. If we find tiger tracks to boot, they become definite stakeout spots.

"Entrance to west slope of Crow Mountain. Fifteen hectares of nut pines. Over three pinecones per branch. Tracks and droppings of unidentified tiger a month later."

These snippets of information come together to create a tiger eco-map, which typically includes the following information: traces of tigers (pugmarks, droppings, claw marks, and the identity of the tiger as deduced through the evidence), traces of tiger hunts (prey, time and place of hunt, where the prey was eaten), traces of ungulates (the species and population of ungulates that traveled through an area, migration routes, and their purpose of visit), and food yield (the location and size of oak, nut pine, and wild walnut trees; how much fruit they bore that year).

Once the eco-map is ready, we can predict to some extent which areas the tigers frequent and which paths they like to take. Based on this information, we decide stakeout spots for the winter and make a more detailed eco-map of the areas we have chosen. The detailed eco-maps contain specific routes that tigers and ungulates take, which help us decide the precise location of our stakeout bunkers, camera angles, and the background that the cameras will capture.

WE WALKED ALONG Deer River. It was the dry season, and the water level was low. The river was about four or five meters wide where the current was moving and seven to eight meters wide in the swamp where the water slowed. At the river bend, a large nut pine uprooted during the flood had made a natural bridge. Tigers and humans alike use these log bridges to cross the river.

Halfway across, I sat on the log and looked down into the water. Spotted fish flapped their fins in every shallow where the current was gentle. The Manchurian trout were in the middle of spawning season. In Ussuri, they spawn between April and May when the river begins to dry. It's so hard to find Manchurian trout in Korea that their habitats are designated Natural Monuments, but they thrive in Ussuri, as far as the eye can see.

The Manchurian trout gently waved their tail fins to and from as they swam against the current. The propulsion of the fins and the force of the current were in perfect balance, so the trout either held against the current in the same spot or slowly made their way upstream. Black dots and broad red vertical stripes ran down the length of their bodies, and their gills were iridescent with a hint of green. The beautifully decorated males surveyed the situation in the water. If a female approached, the male would greet her and escort her back to a safe spot. If a male approached, it would dart at him to chase him away. The trout disappeared and materialized again as they climbed back and forth over the clear, endless ripples. Just as one cannot move against the flow of time, the trout eventually stopped swimming against the current, rode the river down to where they'd begun, and started climbing back up again.

I crossed the log bridge and carefully put down my backpack on the sandbank. Thanks to the long march through the forest, I was sweating as if it were already summer. I took off my shoes and carefully waded into the water, and the Manchurian trout swam away. The riverbed was a mix of sand and small pebbles, ideal for making nests and laying eggs. Every spawning ground was full of yellow eggs.

The water came up to my knees, and the current was neither fast nor slow. I waded in a little deeper. Even at the deepest part, the water came up only to my waist. I approached, clawing at the water, and the fish panicked and scattered instantly. They hid under crimson leaves on the riverbed or under rocks. The water was so cold, my calves started to ache. I washed the sweat off my face and waded out.

Just as I was about to set foot on the riverbank, I froze. There was a pugmark on the ground. They say people are often blind to things right under their noses, and here I was, about to step in a pugmark. I couldn't decipher the exact size of the paw because the pugmark was printed in soft sand, but it must have been made recently, since the outline was still clear.

My heart skipped a beat. I followed the tracks and saw what I hadn't noticed before. Tracks were everywhere. The little strip of sandbank that led upstream was dizzying with tracks. Traces of tigers running, wrestling each other, and lying around. And it wasn't just one tiger. At least three or four tigers had rested and played here.

Tigers are solitary animals that don't enjoy the company of other tigers, yet here was proof that a group of them had frolicked on the sandbank by this clear river. My heart beat wildly. The river lay at the center of Bloody Mary's territory, which meant there was a good chance that the tigers were Bloody Mary's third litter. They weren't cubs anymore, but nearly fully grown tigers.

And then I saw something strange on the ground. Fish tails were strewn about in the grass at the edge of the sandbank. I looked closer and saw they were Manchurian trout tails. I didn't see any heads or bones. It seemed the tigers had eaten the fish that had swum out to shallow waters to lay eggs. I'd heard rumors about tigers catching and eating fish, but this was the first evidence I'd seen with my own eyes.

The best time for tigers to fish is in the fall when salmon and trout travel from the sea to the rivers to lay eggs. In September and October, bears gather along rivers to catch the fish. Tigers also come down to the river during this period for the salmon and the bears, although the bears are the tigers' primary target.

Since fish aren't a staple for tigers, one rarely finds remnants of a tiger's fish feast, let alone sees a tiger catching fish. What's even stranger about what I saw was that the tigers had caught and eaten not salmon in the fall, but Manchurian trout in the spring. Only once in my life had I heard of tigers fishing Manchurian trout, and that was from a ginseng gatherer I am friends with in Mayak Village named Olga Kimonko. She told me a story she'd heard from her father, who was a hunter.

As he had recounted it to Olga, "One year, I was traveling up along the river to hunt a Manchurian red deer. I was going around a sharp river bend when I saw Amba. Amba was just standing there in the middle of the river. I quickly hid behind a rock and waited for Amba to move on. But Amba just stood there for a long time. I looked closer and saw that Amba was staring into the water with one paw up. He stood very still for a long time and then suddenly slapped the paw above the water against the leg that was in the water. It made a big splash. Then Amba pulled something out of the water. It was a Manchurian trout about the size of a deer tail. He pulled it to land and started eating it head first. Amba ate it all and then went back in the water and stood very still again. Isn't that clever?"

I'd once researched an area where toads gather to lay eggs. When I returned some time later, tadpoles had hatched out of the eggs. The tadpoles scattered when I dipped my hand in the water, but soon returned and began to cling to my hand. When I slowly lifted my hand out of the water, I was able to scoop up a handful of tadpoles that were busy nibbling on the dead skin cells of my hand. I thought of the tadpoles when Olga told me the story about what her father saw.

A tiger walked into the middle of the stream. The frightened trout scattered. The tiger lifted one paw and stood still. The water was shallow because of the spring dry spell, but there was enough to make the tiger's fur dance in the current. He waited patiently for the trout to return, one by one. They tapped his legs with their snouts. They plucked and ate the tiger fur, thinking it was an aquatic plant or insect. The tiger watched carefully, then slapped the water, its paw coming down on the fish like a hammer. The tiger waded out of the water with the unconscious fish between its teeth.

It was ingenious of the tiger to use its own leg hair as bait. The method suited female tigers that are patient and cunning rather than powerful. It especially suited Bloody Mary's personality. A silly tiger cub would have splashed around in the river, pouncing on every fish in sight like a bear catching salmon. But that method works only in the fall when the large, slow salmon swim up the river in big schools. It would be impossible to use such a clumsy method to capture the small, impossibly quick Manchurian trout that are far fewer in number.

The traces of Bloody Mary's meal revealed further evidence of her great intelligence, apart from her brilliant trout fishing: she knew when and where the Manchurian trout laid eggs, and she was aware that the spawning season falls during a dry spell when it's easy to go in the water.

The paw prints traveled up to meet Mizinets River. Not very far along, I came across a brook that cut through the forest and merged with the river. On the stretch of mud by the brook, I saw the clean paw prints of the tiger family. Four in total. One of them had a front paw pad width of 9.7 centimeters. Sure enough, it was Bloody Mary. The very same tiger that had left the paw prints I'd found and measured at Tachinko Beach some time ago. She'd had a litter of three: two females, one male. The two female cubs had similar front paw pad widths: 9.2 centimeters and 9.4 centimeters. The other cub's measurement was 10.8 centimeters—a boy.

Female cubs don't become as big as their mothers until they're two years old, but male cubs grow much faster. When a male cub is one year old, his stature and paw pad width are already that of his mother (but his body is still mostly baby fat). The cubs' measurements revealed that they were nearly grown, about a year and a half old. Among the female tigers in Ussuri, Bloody Mary was one of the best mothers. She had once again managed to safely raise three cubs.

In the mating season, female tigers spread their scent to notify male tigers. The scent is so strong and seductive that the male tigers put off hunting, the most important task for tigers, to attend to the females. Almost every male tiger in the territory picks up the scent and comes to a female in heat. The Great King is no exception. And when he turns up, it doesn't matter how many other tigers are also interested—the female mates with the Great King. Based on the timing, Bloody Mary's first and second litters were progeny of the last Great King, whom I'd nicknamed "Tail." But there was a good chance that this litter was fathered by Khajain, the current Great King of Lazovsky and the spirit of Sikhote-Alin.

The fog spread and gradually thickened into a misty rain. It began to drizzle. A raindrop bounced off a long leaf bent back like an orchid and spattered into dozens of little beads. On the faraway Deer River, the delicate lines of drizzle danced in the wind. The spring monsoon had begun. Bloody Mary and her family must have been walking along a river that was swelling and growing ever more enchanting. The traces peppered behind would be washed away in the spring showers.

PART II

Amba, God of the Forest

CHAPTER 5

A Tiger Leaves His Mark

IN MID-JUNE, WE began research for our second expedition. We planned to look around the valley called Dipiko first and then move on to nearby Santago Valley. We would travel along Santago River, over the Dragon Spine, up Shauka River, and down again. Dr. Galina Salkina, a tiger researcher and good friend, had decided to accompany Valosia (her husband) and me on the trip, which would extend over seventy kilometers. The other team would survey the regions up Shauka River, traveling upstream.

Santago River weaves through many parts of the lowlands of the Sikhote-Alin and cuts through the middle of Lazovsky. The deep valley formed along the river is Santago, the estimated northern border of Bloody Mary's territory. Between Santago and a prominent part of the mountain ridge is Dipiko. Both Santago and Dipiko were named by the Udege. The "go" of *Santago* and "ko" of *Dipiko* have different pronunciations, but both mean "large valley." *Dipiko* means "the valley frequented by hunters," which reveals that the valley was once populated by many animals.

These two inland valleys have healthy nut pine and oak forests. The dense oak forest in Dipiko is an especially rare, vibrant forest.

Most of it is made up of Mongolian oaks, and they had produced a great harvest of acorns that year.

On the Dipiko mountain ridge was an unusual area where a large puddle had formed. I had found many wild boar tracks around the puddle recently. The wild boars had taken a mud bath in the puddle and then rubbed their shoulders against nearby oak trunks. Some mud traces on the trunks came all the way up to my chest.

Walking along the forest path, I came across more tracks. They looked similar to tiger cub tracks. Many sets of prints were intermingled. I looked closer and saw that the paw width was small and that there were clear claw marks above the toes. The tracks actually belonged to animals in the Canidae family.

Animal tracks from the cat family and dog family are so similar that it's hard for the untrained eye to tell them apart. But if you look carefully, there are four crucial differences.

First, claw marks. Animals in the cat family retract their sickle-shaped claws when they walk and extend them only when they are digging them into their prey or climbing steep hills full of rocks. Because they extend their claws only when they need them, the claws don't show up on their prints. Animals in the dog family, on the other hand, cannot retract their claws, so claw marks show up in their prints every time.

Second, the size and shape of the paw pad. The prints of an animal in the cat family will always be wider than the prints of a similarly sized animal in the dog family. People often mistake large dog prints for tiger prints simply based on the size. This is because dog prints tend to look large thanks to dogs' more spread-out toes. But if you look carefully, you'll see that, for example, a fully grown mastiff's paw pad is far narrower than the paw pad of a tiger cub. In addition, a cat's paw pad is shaped like a trapezoid, while a dog's is like a triangle.

Third, the length-to-width ratio of the print. Cat tracks are circular, overall, with more or less equal length to width, but dog tracks are more oval, with length greater than width. Foxes have the longest print, followed by wolves, and then dogs. Tigers' back paws also make prints that are slightly longer than they are wide, but their front paws leave prints that have an equal length-to-width ratio or sometimes even greater width than length. Some people call tiger prints "apricot flower prints" because the four toes and paw together resemble an apricot blossom.

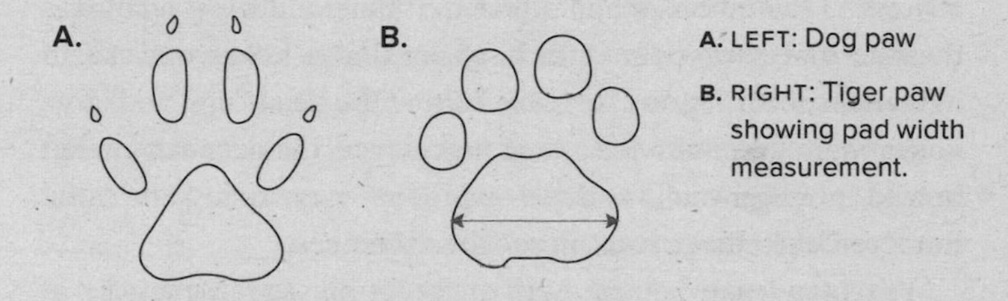

A. LEFT: Dog paw

B. RIGHT: Tiger paw showing pad width measurement.

Last, the position of the toes. In tiger prints, the second toe is positioned the highest, followed by the third, first, and then the fourth. Researchers look at the second toe to tell if the print comes from the right paw or the left. If the toe in the highest position is the second from the left, you're looking at the right paw; if it's second from the right, you're looking at the left paw. The second and third toes of an animal from the dog family, however, are at the same height. This makes it difficult to tell if the print comes from the left paw or right. Also, tiger toes are closer to the ball of the paws, giving them an organized, drawn-together look, unlike the toes of canines, which splay out.

We followed the prints and found a deer carcass. Its head and legs were gone, and many bites were taken from the remaining carcass. This was typical of a canine hunt. Canines hunt in groups and tear at the prey all over. The tiger, however, hunts alone and kills its prey in one bite by snapping the back of its neck. If the prey is too big to kill by snapping its neck in one bite (like a large stag, a wild boar, or a bear), the tiger suffocates it by clamping down on the windpipe. And so the carcasses of tiger kills have only four fang marks on the front or back of the neck. This deer was killed by dogs that had left their human masters and grown wild. An increasing number of Ussuri deer were being hunted by wild dogs lately, some of them drowning or falling to their deaths in their attempts to escape.

Wild dogs also hunt ungulates, but tigers and poachers usually get to them first. When a poacher hunts a deer, it just takes the meat and leaves behind the head, legs, large bones, and skin to lighten the load. Tigers also leave behind the head, legs, and large bones when they hunt deer, so at first glance, the deer carcass left behind by a tiger and a poacher may seem more or less the same. But if you look closer, you can see the difference.

Poachers leave behind human footprints and the tracks of hunting dogs, along with bullet casings and jeep tire tracks. Tigers leave behind their own set of tracks and a trail of the animal being dragged. If the tracks have disappeared, look at the carcass. Poachers leave knife marks on a deer's skin and bones because they skin the deer whole. Tigers, on the other hand, pull out all the fur with their front teeth and then eat the skin as well. Also, while poachers leave behind all the deer's internal organs, tigers eat the organs and leave behind the contents of the deer stomach and intestines. Poachers leave the deer head intact, including the scalp, but an especially hungry tiger sometimes takes a bite out of the deer's cheeks, the back of the head, and the scalp, leaving just the skull. The carcass left behind gives important clues to the cause of the deer's demise.

Nature includes both living spirits and the traces they leave, as well as all the intuitions that surround them. When I say traces,

I'm referring to physical evidence that living things like tigers leave: pugmarks, droppings, claw marks, and other such signs. By intuitions, I don't mean just assumptions and feelings. I'm referring to objective intuitions that come from observing nature for a long time, in the way American Indians or native Ussuri have done. In the forest, intuition is science. When trying to explore a logical path in nature, we come across many forks in the road. We must pause at each and consider it. Collecting all our intuitions and adding up all the physical evidence we find helps us arrive at a scientific decision.

All living things that move leave traces created by their mass pressing on the surface of the planet. This is a law of physics and an important record of life on earth. If we look at these records, we can learn a great deal from them. Traces help us deduce the past and predict the future. The larger the animals, the more we can learn from the traces they leave behind—where they like to go, what they do there, why they do it, and so on. We can also determine their sex and age range, thereby identifying individual tigers.

To identify an individual tiger, it's easier to first divide by gender and then into categories such as juveniles, sub-adults, and adults. Tigers under the age of one are juveniles, aged one to three are sub-adults, and over three are adults. Sub-adult tigers are roughly the same size as adult tigers, making it difficult to tell them apart just by looking at them. Sub-adults are large but not completely independent, which means they have less experience. They're also sexually inactive and mentally immature.

The most important factor in telling individual tigers apart is their prints, specifically the width of their paw pads. This is because tigers have more highly evolved paw pads than other animals do. The width of the front paw pads is especially telling. Tigers' front paw pads are larger and have a greater width-to-length ratio than their hind ones.

A mature male tiger has an average paw pad width of ten and a half to thirteen centimeters. The widest, thirteen centimeters, usually belongs to the Great King in his prime. There are past records of wider paw pads, but hardly any paw pads have reached even thirteen centimeters in recent years.

A mature female tiger has an average paw pad width of eight and a half to ten centimeters. An unofficial record shows that a female tiger paw pad can be up to eleven centimeters wide, but within the small population we have today, the widths hardly ever reach ten centimeters.

If the front paw pad width is less than eight centimeters, the print definitely belongs to a juvenile, and if the width exceeds ten and a half, it's definitely an adult male. But if the measurements are anything from eight and a half to ten centimeters, it could be a mature female or a sub-adult male. With these cases, we determine the sex by factoring in their stride and the length of their bodies. Because tigers' paw pads develop faster than their bodies, a sub-adult tiger's paw pads may be roughly the same as an adult's, but its steps and body length will be shorter.

I got chills as I followed the fresh tiger tracks in the snowy forest. The chills were not like the ones brought on by the cold outside, but more like a psychological coldness that came from deep within. As I followed the track with my heart beating wildly, I came up on signs of the tiger frolicking and resting in the snow. When I saw the clear imprints of a tiger who had been innocently rolling around and playing, my frantic heart finally calmed. It gave me great comfort to know that tigers like to have fun and bond with each other.

Sometimes I see the traces tigers leave by crouching gingerly in the snow. Tigers often sit in this sphinx-like position—hind legs tucked in, front paws stretched out neatly in front—right before they pounce on prey or when they are on alert. The sphinx position with the tiger's hips rotated to the side is called a half-sphinx. A tiger lies in the half-sphinx position when it's resting. If there is ice in the spot where the tiger was lying, it means the snow melted from the tiger's body heat and froze again when it left; in other words, the tiger had taken a long break or a nap in that spot.

We can use these traces to measure the length of a tiger's body. When measuring body lengths using traces of sphinx or half-sphinx positions, we use the length between the base of the tiger's tail and the chest. We do not include parts that can be tucked in, such as the legs or the tail, as part of the body length. If we find a trace where the tiger had lain down completely, we measure from its head to its bottom.

According to *The Terrestrial Mammals of the Far East of USSR* (1984), an adult male tiger's average body length (from head to tip of tail) is 3 meters and its average body weight 200 kilograms. The largest ever recorded, in the *Russian Red Book* (1987), was 4.17 meters long and weighed 350 kilograms. The Bengal tigers that live in India, Nepal, or Bangladesh, on the other hand, are on average 2.5 meters long and weigh 150 kilograms. The harsh climate and environment made the Siberian tiger evolve in different ways than tigers in tropical regions. To reduce body heat loss, the Siberian tiger became larger to reduce its volume-to-surface ratio.

SPHINX POSITION

Average adult female tiger: 120 cm

Average adult male tiger: 145 cm

HALF-SPHINX POSITION

Average adult female tiger: 115 cm

Average adult male tiger: 140 cm

REPOSE POSITION

Average adult female tiger: 160 cm

Average adult male tiger: 195 cm

The same tiger paw can leave different prints depending on the condition of the surface the tiger walks on. Prints on the right kind of mud are accurate representations of the paw. But if the mud contains too much water, the mud slides down as the paw is lifted and makes the print smaller. Dry sand makes prints with large overall outlines because the sand is blown off on impact, but the size of the paw print itself shrinks when sand slides down as the paw is lifted. Prints on fluffy snow are larger, because the snowflakes are also blown off on impact as the paw lands. Prints left after a heavy snow are difficult even to recognize, for the snow collapses as the paw is pulled up. We measure the depth and width of the hole to guess the paw size, but it's difficult to get an exact measurement.

We must also pay close attention to the surface of the print. Is the mud jutting out between the toes wet or dry? Is it well defined or worn down? These questions help us determine when the tiger was there. If there is water in the print, is the water murky or clear? If the former, the tiger went by less than an hour ago; if the latter, it's been more than an hour. If there are ripples in the water, the tiger went by minutes ago. If the print is on grass, we look at which direction the grass is leaning, how wilted the stems are that have snapped under its paws, and, if it's morning, whether dew is clinging to the blades of grass. A snapped branch, a rock out of place, a tuft of hair stuck in a tree trunk, and other things in nature that are out of place all demand our careful attention.

Following a trace is the act of taking one step toward specific from vague. A vague trace becomes clearer with every observation and investigation until it becomes a small but vivid truth as real as something I see before my very eyes. One fact connects to another, and these small facts come together to form the whole truth. Following traces, therefore, is an act of outlining a living thing's way of existence.

All traces disappear with time. The wind removes them, the rain washes them away, and the snow covers all. The cleaning crew of the forest takes apart the carcasses of living things, and time silently erases everything, including the traces of seasons.

Bones, however, last a while. Other things may soon erode, but bones maintain their shape for much longer. Bones are the history of a forest and, in their own way, immortal. Just as you may be moved when you read a classic novel, I am inspired by the history of the forest I read in bones. Traces are physical things that sometimes have spiritual properties that reverberate in our souls. And so the traces I find always remain in my heart.

CHAPTER 6

Temu, the Spirit of Storm

SUMMER WOKE THE forest. The last dull echoes of the wild mallards had died out, and bush warblers and wrens wove the canopy with song. The flowers that had laid colorful carpets in the spring had vanished, and summer blossoms with tall stigmas had taken their place. Red and yellow blooms lined the brooks. Blood irises lured wild bees and insects with their fluttering purple banners.

We had surveyed Dipiko for over a week without finding anything out of the ordinary other than a good number of acorns on the oaks and evidence of the wild boars' exploits. Then we traveled over to Santago and surveyed the river for a week as we made our way upstream. No signs of tigers there either. So we headed toward the Dragon Spine via Urine Rock. Urine Rock is geographically significant because it's on the crossroads between Santago to the west, Dipiko to the south, and the Dragon Spine to the east. Tigers often make an appearance at this intersection and leave their mark by spraying their urine there—hence the name. It's an important landmark for them.

Valosia and Galina, who were walking ahead of me, bent down to study the ground. Tiger prints, clear on the damp ground after the spring showers. Even at a glance, I could tell they were huge. There was probably only one tiger in Lazovsky with a paw pad this size: Khajain. Before Khajain took over Lazovsky, his prints had usually been found in the south of the region. This meant he'd been born there. It was possible he was related to Bloody Mary, whose territory was in southeastern Lazovsky. It was unlikely that they had a parent-child relationship because there wasn't enough of an age difference, but they might have been siblings or half-siblings. They were both over ten years old and in their prime.

Khajain's footsteps took a detour before Urine Rock and led to a birch by the mountain path. There were claw marks on the thick tree trunk. It seemed he had dug his claws in as high as he could on the trunk and then dragged them down. The claw marks were over three meters high. I could barely reach their height by standing up straight and raising my walking stick above my head. Tigers compete with one another by leaving the thickest claw marks on the highest spot. Most tigers leave their marks at about one and a half to two and a half meters, and the Great King is the only tiger who can leave a claw mark at the highest point. Above the base of the trunk was a black stain from years and years of tiger urine. There was some yellowish hair on the black mark. Khajain had made his claw mark and then rubbed his face and neck against the urine stain to cover the scent of other tigers with his own. Tiger hair has a unique scent that can also be used to mark territory in spots other animals and tigers frequent.

The number of territory markers one tiger leaves is far greater than you might think. We have yet to discover the correlation between the number of tigers in a given region and the number of territory markers they leave, but it's clear that the more tigers that use a certain path, the more marks each of them leave. If one tiger leaves a territory marker, this triggers a competitive spirit in other tigers who also rush to leave a mark.

After marking his territory on the birch, Khajain had walked along the mountain path. About five hundred meters ahead of us was a small hole dug into the side of the trail. It was a tiger stamp, a hole in the ground with urine or excrement on the pile of dirt that was dug up, used to mark territory. The tiger digs the earth with its hind legs, and if it can't leave urine or excrement, it just leaves the hole. These holes are usually small, but I once saw one that was two meters long and seventy centimeters wide. The size of the stamp has no relation to the size of the tiger, but has to do with the tiger's personality. Some tigers leave unremarkable stamps that are no more than a few perfunctory scratches, while others habitually leave long, wide stamps. Khajain's stamp was only fifty centimeters long, but so deep that the roots of nearby trees and shrubs were exposed. This was Khajain's trademark.

The excrement and urine in a stamp contain chemicals unique to individual tigers. To the human nose it just smells bad, but tigers can use these scents to tell each other apart. Dogs can, too. Taking advantage of these properties, researchers have been collecting tiger droppings since the 1990s. With the help of well-trained dogs, they divide these samples by smell. As the sample size grows, it gradually becomes possible for dogs to identify each tiger according to dropping smell. This method is very useful in determining the number of tigers that inhabit a certain region and estimating the size of their territory.

We continued to follow Khajain's tracks and found ourselves at Urine Rock, roughly the size of a cottage. The tracks stopped. There was a strong stench coming from a spot on the rock about one and a half meters aboveground. Khajain had lifted his tail here, aimed at the rock, and sprayed his urine. It was pungent. I was certain he had sprayed after the spring showers. A tiger sprays its urine anywhere other tigers can find it—trees, rocks, logs, anywhere. But Urine Rock is special. The rock is overwhelmingly tall and clearly visible from far away. The smell of tiger urine is strong and pungent, so from here it reaches other tigers easily. Deer and wild boars also know what the scent means and stay away.

Once Khajain had sprayed Urine Rock, he had climbed the Dragon Spine. He must have been busy reinforcing the territory markers that had washed away in the spring showers. It's the duty of the Great King to leave his scent, claw marks, and excrement all over Lazovsky to declare to beasts and humans that this land is his. Tail, Khajain's predecessor, had also come by Urine Rock frequently after snow or rain.

I took out the plaster I had prepared and added water from the stream to form a thick mixture. I poured the mixture into Khajain's front paw print. Twenty minutes later, I lifted out the hardened plaster and dusted off the dirt. The plaster mold of the Great King's paw was complete. It was evidence of Khajain's existence and a part of the history of Ussuri forest. I would place this next to the plaster molds of his predecessors.

After sundown, we pitched our tents by the stream and started a fire. Over dinner, Galina brought up a story about Khajain.

"We have a record of Khajain traveling ninety-two kilometers in fifteen hours. It was a coincidence that we were able to record it. A forest ranger in northern Lazovsky happened to pick up a fresh dropping, and the dogs confirmed that it was Khajain's. But then the next day, another forest ranger from southern Lazovsky brought in another fresh dropping, and it was also Khajain's. The elapsed time was fifteen hours. He'd managed to travel ninety-two kilometers in that time."

It is very rare for tigers to move in a straight line so quickly. And even if the tiger really does it, it's difficult to measure the time and distance it travels. Tigers generally travel at a leisurely pace. If prey is abundant, the tiger stays for a while, but if it isn't, it leaves. There are many places tigers stop along the way—places to mark territory, places they've hunted before, places suitable for rest, the place they were born, and so on. Even if the tiger is traveling a mere ninety-two kilometers, the trip could take a month or two.

But if a male picks up the scent of a female in heat or a mother is looking for her cubs, it's a different story. When such urgent matters come up, tigers forget everything else and head to their destination in a straight line. But they don't run at full speed like they do when hunting. Carnivores like tigers are good short-distance runners, but lack the endurance to run long distances. They get tired after running just a few kilometers at full speed, so they jog at a steady pace instead. This way, tigers can cover a distance of close to one hundred kilometers over twenty-four hours. I'll bet they could go faster if they wanted to.

We'd planned to climb the Dragon Spine the next day, the first of July, the beginning of the second half of the year. As the last day of June wound to a close, the wind was gentle, but a thick plume of cloud was gathering above the mountains. The sun cast its last light through the rolling clouds. The sunset was as red as blood.

The next day came with frenzied wind, which blew in all directions as we started our ascent. Then it began to rain. When we were about halfway up the Dragon Spine, the raindrops became heavier. We gathered branches, placed them on top of a shrub, covered it with a big plastic tarp, and added more branches on top of the tarp to make a roof. If it had been a passing thunderstorm, this should have been enough to keep us dry. But the rain came down harder. The sky opened and water poured down like a waterfall. Every dry crook in the mountain slope overflowed with a stream of rainwater as deep as a meter. A terribly strong wind blew away our roof. The forest swayed and nut pine branches snapped and flew away. Suddenly, there was a burst of light and everything was illuminated. I saw lightning shoot straight down, followed by the crash of thunder. With a monstrous sound, an elm that seemed at least a hundred years old split in two before my eyes. The tree fell slowly, taking the surrounding young trees with it. Trees with large trunks were uprooted and carried away by the stream. A whirlwind sucked up the yellow water of the river and tossed it everywhere. Temu, the spirit of sea and storms, had paid us a visit.

We crawled up the Dragon Spine in the downpour. Once I was as soaked as if I were crawling up a river, I found myself enjoying the rain. Only after we passed the summit did we find a little cave where we could take shelter. It was eerie inside the cave, which wasn't small, but too small to pitch a tent.

I took off my backpack and caught my breath. Valosia and Galina looked like drowned rats. Fortunately, there was a big pile of leaves in the cave. I could make out the vague indentation of a large animal that had been lying on the pile of leaves. I saw yellow hair as well. A tiger must have been here. Was it Khajain who climbed the Dragon Spine ahead of us? I saw the paw prints of a badger, too. It made me feel closer to the poor creature as I pictured it huddled here to keep dry. Animals and humans alike are helpless in bad storms.

We gathered a pile of leaves and lit a fire to dry ourselves. The smoke filled the cave, making our eyes red. As the air slowly cleared, the cave grew warm. We collected rainwater for tea and had a brief snack. Valosia lit his cigarette on the campfire, and Galina scolded him for stinking up the tiger cave with cigarette smoke. Valosia seemed nevertheless happy as a child. We discussed this and that before we all fell silent again. We listened to the sound of the wind threatening to blow everything off the face of the earth and the rain pouring down as if there were a hole in the sky. It didn't seem like the storm would let up any time soon.

When I live in the forest, I become absorbed by the various sensations of nature—the pure morning air, the swamp-like morning fog, the still noontime sunlight, the soft evening breeze, the blue cold of the frozen taiga, and the white snow flowers of the dazzling snowy landscape. I watched the majestic storm through the opening of the cave. What would the cave look like from the storm outside? Peace in the midst of chaos!

I piled up some leaves and unrolled my mattress on top of them, then crawled into my sleeping bag. The sound of the storm lulled me to sleep. Which tiger was it that had lain in this very spot where I was lying now? Where had Bloody Mary's family found shelter in this storm? My imagination ran away on me, mingling with the chaos of outside, and faded as I drifted off to sleep.

CHAPTER 7

Deer Hunting in the Fog

THE STORM RAGED on for two days, but started to let up by dawn of the third day, and the rain finally stopped that morning. Rays of sunlight streamed down from the east, and the sky cleared. Enduri had calmed the wild temper of Temu. I walked outside the cave and looked down at the forest. The cloud was chasing after the tail of the storm. The forest had become thinner, but the air was clean and a mist hung above the river. In the canopies of pine trees poking above the mist, a flock of crows was making a racket.

Crows are the chatterboxes of the forest that meddle with every little thing that goes on in their neck of the woods. Where there are crows causing a commotion, there's always something happening. When carnivores have hunted prey, crows somehow find out and spread the news far and wide. They hang around patiently for a little morsel of meat and follow the tigers if they drag their prey elsewhere. For Ussuri tigers who like to move around stealthily, these crows are a nuisance. But for us, the crows were helpful informants. We never failed to check out areas where crows and jays had gathered or where eagles were circling above.

We carefully made our way down the Dragon Spine and along the river. Cascading mud had erased the path and made the riverbank very slippery. The river and forest were inundated with a milky mist. I couldn't see so much as a foot in front of me and felt as though I were in a sea of fog. In the fog, I saw something black moving along lazily. It waded across the river and approached me. My heart pounded. The wind stirred the thick fog and unveiled the mystery: it was a towering Asian black bear.

Perhaps it was the fog or its bad eyesight, but it continued heading toward me even though I was standing right in its path. My heart raced faster. I thought about retreating, but stood still. The bear that had been looking straight at me finally grasped the situation, and his expression changed instantly. He quickly turned and ran down the wild, muddy river and disappeared into the fog again.

It's curious. I never see bears when I'm wandering the forest looking for them, yet in a moment like that, when the forest was in disarray after the storm and all parties were distracted, we had run into each other. It may have had bad eyesight, but bears have good senses of smell and hearing, and it had come up to me all on its own. It had all happened so quickly that I could hardly believe I had just seen a bear except for the fact that I distinctly remembered thinking that bears swim as well as tigers do.

The landscape along the river had changed, and so had the direction of the tributaries. The muddy water rushed past beneath my feet. The water level had risen. The large trees that had been uprooted and carried away by the river were stuck here and there in the riverbed. A few hundred-year-old trees that had washed down, roots and all, had formed a new log bridge. Clumps of dirt and young trees clung to the thick roots jutting above the whirlpool of yellow foam, sending bits downstream in the rushing current. With so many old trees drowned in the river, the spirit of the water must have been overjoyed.

If you look carefully on a clear day, you can see the outlines of old trees and logs under the rippling water of a river that runs through the forest. The movement of the water creates an optical illusion and makes the logs seem like great slithering mythical snake monsters. If the log happens to be coniferous, like a pine or a nut pine, the tortoise shell–like bark looks just like the scale of a giant reptile.

The spirits of the water, whom the Udege worship, live in such places. Runge, a water spirit in the service of Temu, dwells in rotten trees at the bottom of rivers. The Udege, therefore, are careful not to mistreat the sunken logs. If they happen to stab a log while trying to catch a fish with a spear, they ask Runge for forgiveness: "O, spirit of the water! Please forgive me, for I did not mean to harm you!" And when the river is flooded, as it was after the storm we'd just had, they burn dandelions as an offering to the spirits of the water. It's a ritual to stop the water from coming.

We gathered some azaleas and sent them floating down the river, then carefully continued on our way upstream along the forest path. The trail was muddy and the forest was dripping. Torn by the wind and weighed down by the rain, the leaves were no longer as glossy as if they were coated in oil. But the tree canopy was coming back to life bit by bit as the morning sunlight revived it. The birds had begun to sing again.

About two hours later, we found tiger tracks again. They were as clear as if the tiger had just gone by. On the damp ground, there was blood and a trace of something being dragged. It had been hauled out of the woods on the left, then briefly along the path, then into the woods on the right. There was blood on the leaves as well, meaning it had been spilled after the rain had stopped that morning. We measured the width of the front pad: 9.7 centimeters.

The uproarious sound of crows cawing came closer. Tracks of several tigers appeared. The cubs had joined the hunter. It was Bloody Mary. We ventured deeper into the forest and saw frenzied crows flying about in the thick fog. When we got closer, a flock of crows scattered and the outline of the prey became clear. It was a doe. The flesh was still fresh and blood had not congealed yet. Only a quarter of it remained. There was another deer about ten meters away. It was a fawn, maybe one or two months old. All that remained of it were its hooves, bones, and head. Its young eyes, still open, were stricken with fear.

I could see in my mind's eye a clear image of Bloody Mary out for a hunt the moment the storm had passed. She would have used the thick post-storm fog and the wet forest ground that muted the sound of movement to her advantage. The Ussuri sika deer mother and daughter had fallen prey to her cunning. A cub had got the fawn as it tried to get away. Of the three cubs, the female with the smaller paws had killed the fawn a short distance away and dragged it over here to eat it.

The murky water in the muddy tiger track was still moving. Bloody Mary and her family had just been here tearing the flesh off these deer. They must have left when they heard us coming. I looked around, but all I saw was fog. They must have hidden themselves somewhere in the forest and were watching our every move. To allow Bloody Mary and her family to eat in peace, we collected some basic data and headed toward the river, making a lot of noise so that the tigers would know we were leaving. We traveled downstream along the river, where they could see us leave as well.

It would interfere with our work in many ways if Bloody Mary were to see us as people who pursued tigers. It was best to be seen as passers-by who just happened to be near her. It would have been obvious to the tigers that we were up to something if we lingered at one area for a while doing a thorough sweep for data. Bloody Mary would detect the dissimilarity in our behavior and then act differently herself.

Domesticated animals like cats and dogs can look at their human companions' facial expressions and discern their moods and whether the humans like them or not. The same is true for smart tigers in the wild. *Why are those humans here? By coincidence or by design?* They figure out human intentions based on behavior, expressions, and the energy radiated by people and take precautions or even attack accordingly.

A jay once built a nest in the juniper tree at a temple I used to go to. Out of curiosity one day, a monk at the temple peeked inside and happened to meet the gaze of the jay brooding an egg. The monk felt sorry, as if he'd invaded someone's privacy by looking into their bedroom. From that day on, the monk purposefully ignored the jay when he passed by the nest. The jay also grew to ignore the presence of the monk coming and going, and it was able to raise its young and leave the nest. In contrast, an azure-winged magpie once built a nest in my friend's garden. Enchanted by its light blue wings and long tail, my friend looked in on the bird often. Not long after, the magpie gave up the nest and flew away, leaving behind a rotten egg.

We can't see all of nature with our own eyes, and there is no need to. Like the monk who believed that the baby jay would grow safely into adulthood without him watching, it's important that we believe without seeing. If the monk ignores the jay, the jay ignores the monk; if we ignore the tigers, the tigers ignore us, too.

CHAPTER 8

Survival-of-the-Fittest Cub

SPLASH!

The water spontaneously jumped and writhed. It was the taimen, which, at nearly two meters long and one hundred kilograms, is the largest trout in the world. In floods, taimen swim upstream.

A water snake also swam against the current, thrashing from head to tail. Fighting the strong current brought by the flood, it crossed the river with its head held high and its belly pulled as far out of the water as possible. In the water snake's wake, a frog slowly rose to the surface. The frog froze just below the water with only its big, popping eyes above the surface, keeping a watchful eye on the passing water snake. Deciding it was safe, it kicked its legs to get to the edge of the water and slowly climbed a tree trunk submerged by the flood. It seemed tired. At that moment, a maroon creature darted like lightning through the muddy water and shot up toward the tree trunk. The startled frog jumped in the air. The heavy taimen nimbly turned in the air, swallowed the frog whole, and fell back in the river. It drew a long curve on the water with its muscular back and vanished as if nothing had happened.

As the sun sailed to the top of the cloudless sky, the temperature continued to rise. We found a good spot on the river and unloaded our backpacks. To shake off the post-storm fatigue, we had decided to pitch camp earlier in the day. We pulled out everything that was wet—tent, clothes—and left it out to dry. For dinner, we would have taimen prepared in traditional Udege style.

First, we went into the forest to gather mushrooms. The forest was mushroom heaven. The mycelium, or the thread-like vegetative phase of fungi that grows on rotten trees or leaves, had absolutely taken over the moment it stopped raining. After a summer rain, the mycelium grows and forms the shape of a mushroom within six hours. Mushrooms come in a variety of shapes after which they are named—turkey tails, puffballs, rounded earthstars, shaggy ink caps, and long net stinkhorns. They were ballooning in every nook and cranny. The spirit of the earth lives in fallen, rotten trees all over the forest. Mushrooms are a summer gift from the spirit of the earth.

Living in the woods, you come to appreciate mushrooms. They absorb the toxicity of other ingredients or the smell of fish and add a fragrant, clean taste to food. If you learn to identify poisonous varieties like *Inocybe* mushrooms and fly agaric, mushrooms can be a great help. We picked armfuls of pine mushrooms and jelly ears. We also gathered lots of Alpine leeks.

In the afternoon, graylings started popping out of the water to eat mosquitoes or mayflies that flew close to the surface. I cast a line and quickly felt a bite. I caught a few graylings and then used them as bait to catch the taimen.

The sun was beginning to set. Soon it would be time for the taimen to come out in search of dinner. After a flood, taimen tend to gather on the edge of the water rather than in the middle of the river because the riverbank is rich with insects that had drowned as the water level had risen. Since taimen are carnivorous, I kept yanking the line and moving the bait to make it look alive.

About ten minutes later, I felt a tug. I pulled as hard as I could. A large taimen jumped once as I pulled. I saw its bluish-brown back.

It thrashed once more, and the line went slack. It's not easy fishing with a barbless hook. I cast again and gazed at the line, which seductively wavered each time I gave it a tug.

The sun sank below the horizon, casting the long evening light across the surface of the water. The shadow of the mountain furtively crept into the current. Saturated by the colors of the sunset, the river flowed. Insects flew along the current and fish jumped below them. The dark silhouettes of jumping fish danced above the surface and dove back with lively splashes of dark droplet-shaped silhouettes. When it all calmed, the silvery current resumed its journey. I knew that the glow of the water was an optical illusion, but I was captivated by the unhurried flow all the same.

Ping!

The fishing line was yanked into the water and pulled taut. I could feel something at the end of the line, which felt like it was about to snap. I reeled it in slowly as the fishing line drew an arc like a compass.

With a flutter and a splash, the taimen leapt out of the water and then fell again. It was a big one. I quickly reeled in the line to maintain tension and then slowed down, giving the taimen time to exhaust itself. After about ten minutes, I started to reel in harder. The large trout drew a neat wake on the water as I pulled it in. Near the riverbank, the taimen struggled once more in vain, tossing its muscular body. I kept the line taut, remembering that I'd lost the first fish by letting the line hang loose at this point. The taimen struggled once more when its head was out of the water, but I dragged its heft to land. It jumped and rolled around on the sand; before long it lay still with its mouth opening and closing.

Living as we did off of what the forest provided, I felt grateful for the generosity of nature. In my head, I thanked Runge.

I gutted the trout and pulled out the entrails. I washed it clean, sprinkled salt on the inside and outside, and stuffed it with leeks and mushrooms. To the side, I spread out several layers of Alpine leeks and then a layer of mushrooms on top. I placed the trout on top of that, and added more mushrooms. I wrapped the trout with the Alpine leeks underneath, placed the bundle on top of the fire's embers, and pressed it down with a flat, wide rock. On the flat rock, I made a fire as well. This is how the Udege cook taimen.

We took out a bottle of spirit that is 99 percent alcohol. We bring a bottle of this strong alcohol with us to limit the weight we have to carry when we head into the mountains. We drew some clean water to dilute the spirit. If consumed without diluting, the alcohol can burn the esophagus. If watered down to 60 percent, it makes *samogon,* a popular Russian moonshine. Forty percent makes vodka.

We poured some spirit on the fire. The flames flared up. The Udege always throw some alcohol or food in the fire when they build one. It's an offering so that Puza, the spirit of fire, will bring them peace and luck.

Puza lives in old trees or hearths where there is phosphorescence. When a woodsman cuts down an old tree, he must first ask Puza to leave. One must not poke around a fire or a hearth with a sharp piece of metal, because it might hurt Puza.

After it was in the hot fire for an hour, we lifted off the top stone and placed the large trout wrapped in Alpine leeks on it. We peeled off the charred leaves to see steam rising from the layer of half-burnt mushrooms. The mushrooms had absorbed just enough fat from the trout to make the meat tender, but not too greasy. The pungent smell of fish had been tempered by the mushroom layer and replaced by the subtle scent of mushrooms and garlic. For the first time in a while, we enjoyed our feast with vodka.

Stars filled the clear night sky. After the storm, the gentle wind of the river felt cool. The flames danced in the breeze. All feelings are purified before a fire, and the tender emotions of times long gone come back to life. We sat and discussed the past and future of Siberian tigers and the groundbreaking tiger researcher L. G. Kaplanov. If it weren't for his efforts and sacrifice, we wouldn't have been able to see the most courageous, most beautiful, and largest carnivore on earth, except in zoos.

The population of wild tigers in Lazovsky has remained between eight and twelve since the late 1990s. Even though the population growth in this area has stagnated, annual reports say that the total population of Siberian tigers has grown each year. In 2010, the American research team at the Sikhote-Alin Nature Reserve reported the total population as approximately five hundred. The number of Siberian tigers surviving in the wild in China and North Korea is under fifty. Between thirty and forty live in Manchuria, ten in the Korean Peninsula, and the rest of the five hundred in Ussuri—but some researchers think that number is inflated.

In Ussuri, the customary way to count tigers is the parallel method. Kaplanov, the first scientist to systematically count tigers, implemented this method, which involves many research teams counting at the same time as they travel along major rivers in tiger territories. Because it investigates every major route that tigers take in the given region, the method has the advantage of getting a relatively precise count of tigers in the area. However, it requires a lot of manpower and a costly budget, because the entire vast region is investigated at the same time. To cut costs, laypersons such as hunters and villagers are recruited to help. Since it's often confusing even for experts to tell if a pugmark belongs to a male or a female, it goes without saying that laypersons can get it wrong. They confuse dog or leopard tracks with tiger tracks or count the same tiger twice when they see tracks in another region. More conservative estimates place roughly three hundred tigers in Ussuri. They're endangered animals, and their population has neither increased nor shrunk. The birth of new tigers does not add to the population count, because the same number of tigers die each year.

They say tigers raise only the fittest cub and abandon the rest. While it's true that only the fittest cubs survive, it isn't because some are abandoned. There's a sad reality behind this claim. Mother tigers try hard to raise all their cubs, but with their habitat shrinking, the number of poachers increasing, and the number of ungulates decreasing, over half the cubs in a litter die of starvation, disease, or accidents before they reach adulthood. Even mature tigers sometimes fall victim to snares and booby traps, but that's nothing compared to the unfavorable odds that cubs are dealt until they become sexually mature and mentally independent at the age of three. Only one or two lucky, strong cubs survive.

Things were worse in the past. Wild tigers were nearly wiped out in Ussuri at one point. In the early 1900s, when Kaplanov was born, an estimated six hundred to eight hundred tigers lived in Ussuri. But the population rapidly plummeted because hunting tigers was legal and guns quickly replaced arrows and spears as the preferred hunting tool. Back then, tiger hunters were considered heroes. It was a time when people who pointed guns at tigers could claim to be "poaching scientists."

Kaplanov grew up watching countless tigers slaughtered. He believed this would lead to no good and decided to devote his life to researching tigers. The more he wandered the forest looking for tigers, the more he realized that their traces were disappearing. And then one day, they vanished altogether. For twenty years, there were no tiger tracks on the eastern slope of the Sikhote-Alin Range, and it seemed no tiger ever visited the Bikin River Valley. Tigers had vanished even in the Lazovsky Nature Reserve. Researchers could not find a single tiger-eaten deer carcass from 1936 to 1948, but there was a staggering rise in the wolf population. Wolves were often seen all over Lazovsky, hunting deer in groups.

Kaplanov felt the need for a more systematic way to count tigers. In 1939 and 1940, he counted tigers by spending the entire winters walking through snow-covered terrain in search of tiger tracks. He concluded that there were no more than thirty tigers in the Ussuri region and that perhaps the number was closer to twenty. Of these, three were shot in 1940.

It came as a shock to Kaplanov that the Siberian tiger, once boasting a population of over ten thousand across Ussuri, Manchuria, and the Korean Peninsula, had only twenty to thirty members left in its species. If he did nothing, they were sure to go extinct. Kaplanov continued his research and prepared a paper on the Siberian tiger's sharp population decline, and he fought to prevent tiger poaching. And then one day, he was murdered by a poacher. He was thirty-two years old.

His paper was published posthumously. In it, he warned that Siberian tigers would soon become extinct. He pleaded for the illegalization of all tiger hunting and a five-year suspension of tiger cub captures for zoos. In 1947, the then Soviet Union passed a law that banned all tiger hunting. In 1956, the five-year suspension on the live capture of tiger cubs, whether for research or for zoos, was also passed.

Just as Kaplanov argued, it was only after the ban on tiger poaching that tiger tracks emerged again in Ussuri. The tiger population in Ussuri slowly grew, and in the early 1970s, the ungulate carcasses left by tigers finally outnumbered ones left by other carnivores. According to a report in 1978, the tiger population had risen to two hundred.

But in the early 1990s, when the Soviet Union began to dissolve, the situation rapidly changed for human society and nature alike, spelling another crisis for the Siberian tiger. In this time of social turmoil, an enormous international market for the selling and buying of all commercially valuable wildlife opened in Ussuri, just a stone's throw away from China, Korea, and Japan, adding fuel to the rising demands.

The most valued commodity was the tiger. This triggered a systematic, grand-scale poaching operation that spread like wildfire.

Special snares, booby traps, and even landmines were invented specifically for tigers. During this period, approximately sixty tigers were illegally hunted every year. This had a direct impact on the tiger population in the nature reserves, which was reduced by more than half in the mid-1990s. The ungulate population also took a nosedive.

The political situation in Russia eventually settled, and international efforts to protect tigers caused their population to stabilize again in the late 1990s. From the twenty or thirty tigers Kaplanov presumed remaining in 1940, the population multiplied tenfold, ushering tigers out of a serious risk of extinction. This number, however, is not enough to guarantee long-term survival. There are still a variety of natural and manmade factors that threaten the species.

If Kaplanov's proposal hadn't been implemented in 1947, the Ussuri tigers would have met the same fate as their Korean and Manchurian counterparts. In commemoration of Kaplanov's forward-thinking efforts and sacrifice, the Lazovsky Nature Reserve was renamed the L. G. Kaplanov Lazovsky National Nature Reserve.

I OPENED MY eyes. I heard the quiet chirps of small birds. When I drew my tent flap, morning dew rolled off the roof. The river flowed beneath the light mist.

On my way down to the river, I stopped abruptly. Tiger tracks. The tracks came out of the forest, stopped in front of Galina's tent, changed course, circled twice the tent where Valosia and I had slept, and then disappeared back into the forest. Bloody Mary had paid us a visit when we were all sleeping. It wasn't surprising given her personality. This was to be expected since we were in her territory, but I was still amazed by her skill and persistence to spy on us without being seen.

The previous night, Bloody Mary had probably left her cubs somewhere and kept watch alone in the forest. She would have waited for us all to fall asleep and cautiously come out to investigate. She had circled Valosia's and my tent twice to see what we were up to and whether we smelled of metal, signaling danger. But all she'd needed to suss out Galina was a quick sniff.

Dunkai, a Nanai, is a hunter and an artist who draws tigers. He once told me that tigers are able to tell people apart. "The Nanai never shoot tigers," he said. "The same goes for the Udege. This is wise for the Nanai and Udege, because if you shoot a tiger, tigers will pursue you for as long as you live. The Nanai always use a walking stick in the forest, so the tigers know from the sound of the walking stick that the man is a Nanai. They also know that Nanais will not harm them, so they aren't wary of anyone carrying a walking stick. But they are suspicious of Russian hunters who carry guns. They stalk them and when they see an opportunity, attack the Russian hunters."

Tigers can tell hunters apart from non-hunters. They know the difference between an herb collector's satchel and a hunter's rifle and can differentiate between a man and a woman by the way they dress and from the smell of cosmetics or cigarettes. They also know that men are far more dangerous than women. This explains why Bloody Mary did not worry about Galina but circled our tent twice.

We would be able to find out more about Bloody Mary if we were to follow her tracks, but she had her cubs with her. A female tiger traveling with cubs has a hair-trigger sensitivity. Given her disposition, she would find us suspicious if we were to follow her then. She may still have been hidden in the forest somewhere, watching us that very moment. We decided not to do anything that could make her nervous.

Leaving Bloody Mary and her family behind, we headed for Shauka. The water level in the river had gone down a great deal.

CHAPTER 9

Tiger Family on the Beach

COOL WINDS BLEW from the East Sea on the coastal cliff of Tachinko. This was the northern border of Bloody Mary's territory. Mayak Village, located about sixty kilometers south along the coast, marked the southern border of her territory. The coastal range that branched out from the Sikhote-Alin Range ran all the way down to Mayak Village.

We traveled down the coastal ridge for several days and came across an old tree standing on one part of the ridge. The circumference of the tree trunk was so large that it would take four adults joining hands to reach around it, and its branches snaking toward the sky formed a dome. White and red pieces of cloth tied to the end of the branches flapped in the wind like wild hair—evidence of people's pilgrimages to a very remote part of the forest. This was Yaloan Tuke, the "Sky Tree of Life."

Tungusic tribes throughout history, including the Nanai and Udege, subscribed to the belief that they were the children of heaven. Each time a country was founded, its founding myth reflected this "children of heaven" theology.

In this theology, a large bird or tree acts as the intermediary between heaven and earth. It operates as a medium. Large mountaintop trees such as Yaloan Tuke were thought to be the Sky Trees of Life or the cosmic trees of shamans and were considered especially sacred.

The energy of Yaloan Tuke is in contact with the sky and the universe. The tree connects the energy of the earth and all that dwell on it with the energy of the universe. Enduri made this tree so that the energy of the middleworld could connect with the energy of the world above and maintain a balance. When the tree was made, a spirit was sent to protect it. The trunk growing out of the spirit's back is firmly attached to the Sky Tree and helps channel the energy back and forth between sky, forest, and earth.

The idea of this tree traveled across Siberia and over the Bering Sea to Alaska and the American continents, where it became the "spirit tree." The spirit tree is represented as the Tree of Souls in James Cameron's film *Avatar*, where it connects the Na'vi people to their deity. The film depicts an animistic circle of life where people are born in nature, communicate with nature through their lives, and return to nature in the end.

The native Ussuri open their hearts to the Sky Tree of Life. Without sincerity, their spirits cannot connect with the spirit of heaven. If they do make a connection, their illnesses can be healed, they can ask for a baby, and they can divine the future. And each time they pass the tree on their way to hunt or harvest roots, they tie a red or white ribbon on the branches. The ribbons are a sign of respect to Amba, the god of the forest.

Though Yaloan Tuke is a sacred symbol for the native Ussuri, it is also an important territorial landmark for the native tigers. Tigers prefer to mark their scents on larger trees because the larger the tree, the more conspicuous it is to other tigers.

There were tiger claw marks on the trunk of Yaloan Tuke. It also smelled of urine. The Sky Tree was an important landmark for the Great King and for Bloody Mary. They came by often when they were patrolling the costal ridge at the eastern end of their territory.

A SMALL BLACK and white dog came bounding down the beach. I sat on the sand and waited with my arms open. The dog jumped into my embrace and licked my hands and face, her tail wagging excitedly. This was Chara, the "four-eyed" lodge-guarding dog. People called her four-eyed because she had white spots on her forehead that made it look like she had two extra eyes. She had been guarding the lodge farther inland in the north until the year before, but she must have moved here with the forest ranger. She was as cute and sweet as ever, but she was getting old.

There was a mountain lodge at the border where the beach ended and the forest began: the Lazovsky Nature Reserve Petrova Lodge. The forest ranger guarding the lodge greeted us, and we unpacked our things inside. The forest ranger told us he had seen tiger tracks on the path leading to Bishannie two days earlier.

It was roughly fifteen kilometers to Mayak Village, and there were five beaches between the lodge and there, all with Udege names. The Udege don't have a writing system, but the names are sometimes written phonetically.

The next day, we set out for Bishannie. As the forest ranger had said, there were tiger tracks on the mountain path. There were two tigers, and based on paw pad width, it seemed they were the male and female cubs of Bloody Mary. We had seen their tracks inland a month earlier on the outskirts of the Dragon Spine. They must have been patrolling their territory. Bloody Mary's tracks weren't around, but she wasn't far away.

Cubs around the age of two don't always travel with their mothers. Cubs usually become semi-independent at around a year and a half old. They're on their own for a few days and then meet up with their mothers before parting ways again. Like hunting dogs that leave their owners to travel and hunt, they eventually return. In this phase, they eat some animals their mothers have hunted and hunt other large animals like deer and wild boars on their own, preparing to become independent. The time when they become completely independent depends on the availability of prey in their territories, but it's usually at the two-and-a-half-year point when the mother starts to breed again.

The siblings' tracks led to the beach, which was isolated and undisturbed by the tracks of other animals. The waves washed up and down the fine, silky sand. If I were an Udege, I would have named this beach Bishannie (soft sand), too. The two tigers had walked alongside each other down the beach, leaving two long trails behind. Fine sand with a little bit of moisture allowed for clean tracks. They were beautiful. I always like the feeling that tracks like those ones give me—the fresh energy of live animals.

The forest ranger had seen the tigers' tracks three days earlier, and high tide at full moon had been four days before that. When it's high tide at full moon, the moon's gravitational pull makes the tidal range the greatest. So for two to three days before and after the full moon high tide, great waves wash everything from the sand. Therefore, Bloody Mary's cubs had been on the beach three or four days earlier, after the full moon high tide, at night. Tigers almost never show themselves in such wide-open places as this in broad daylight.

A large tree that had been ripped from its place in the forest and dragged out here long ago was lying on the beach. Half the tree was in the water, and the roots, the length of a grown man, were buried in the sand. Bathed in saltwater for years, the tree had bleached but was not rotten. On the roots of the great old tree, spreading out like a bushy beard, the tigers had left their marks. They had scratched and urinated. It must have been the cubs. An experienced tiger like Bloody Mary would not have marked her territory in a place such as this, which is wiped clean with each full moon. The cubs must have been practicing. But rather than making playful little scratches, they had left long vertical furrows similar to their mother's. They were starting to prepare for their independence.

The coastal cliff grew more perilous with every peak we climbed, then suddenly, the view opened up before us. Jagged rocks of all shapes and sizes were strewn at the bottom of the dizzying cliff. This was Malaya Apasna (small but dangerous) Beach. Past Malaya Apasna and a small cape, great waves crashed on the base of the cliff and stately rocks occupied another narrow beach. It would be difficult to dock a tugboat there, as there was hardly any sand. It was beautiful but potentially dangerous. This was Apasna (dangerous) Beach.

We climbed carefully down to the beach. There were tiger tracks there—from three tigers. Bloody Mary had joined the two cubs that left tracks on the sand. But the third cub was nowhere to be found. Had something happened to her? Probably not. Two is nearly the age of independence. If a two-year-old cub were to be separated from its mother, it would be able to survive. If nothing had happened, the absent cub had matured faster than the other two. The closer cubs are to independence, the farther they wander from their mothers, and for longer periods.

It's easy for tigers that have parted to meet again. Their keen senses of smell and hearing combined with the habit of continuously marking and leaving their scent help them track each other down in no time. If they're looking for family, it's even easier. Like people, tigers are intimately familiar with the personality and habits of other members of their family. Their thorough knowledge of the forest, where they've hunted and played since they were cubs, also helps them track down their family members. Compared to tigers, we don't know the first thing about tracking research.

The mother and cubs had rested by a rock on the beach. There were marks in the sand where they had rolled from side to side, like the impression that a rolling barrel would make. Bloody Mary's family had played and frolicked and had a good time. They would have licked each other and watched as ships sailed by. They had not forgotten to urinate on the rock before they left.

After their break, the family had headed south along the coastal ridge for Mayak Village. On the way, the path to Triparashonka (three little boars) Beach was forbidding. It was riddled with cliffs and rocky peaks, and large pines, nut pines, and oaks had sprung up in between. The view of the East Sea from the top of the peak was refreshing.

The area between the mountaintop and Triparashonka Beach is called Azalea Cliff. There is a good field of azaleas among the coniferous and deciduous trees. In the spring, the entire cliff turns pink. In the middle of Azalea Cliff are two caves where bears hibernate. Sunny because their entrances face southeast, these deep caves are occupied by bears every winter.

Triparashonka Beach is a pebble beach, rare for this region. The beach is composed almost entirely of blue-black basalt and has hardly any sand. Washed in the waves for an eternity, each pebble is shaped like an egg and glittery. I'm not sure why the Udege named this beach Triparashonka. I once asked Olga Kimonko, the Udege woman I know from Mayak Village, if it had anything to do with the shape of the pebbles on the beach resembling wild boar droppings. Perhaps there's a legend somewhere about three little pigs overcoming hardship and living happily ever after on this beach.

Bloody Mary had passed through Triparashonka Beach and straight on to Diplyak (warm even in the winter) Beach. Diplyak is a cozy basin with mountains all around it. As its name suggests, the Diplyak Basin is warm in the winter because it faces south and gets good light. The coastal ridge is not so forbidding that mountain goats cannot live on it, but it resembles Tachinko overall. The basin has a dense population of oaks. This oak forest, which begins at Triparashonka Beach, unfurls all the way to Mayak Village.

We entered the forest. The oak canopy nearly blocked the sky. The shadows of oak leaves danced on the forest ground, and splotches of clear sunlight danced between them. Acorns that had ripened handsomely in the summer sun hung from branches in dense clusters. September was drawing near, and some of the acorns had already fallen off the trees.

Acorns plopped to the ground, breaking the stillness in the forest. Acorns are the fuel of nature. They fall to the earth and become food for worms, which become breakfast for birds. Squirrels and gerbils eat some and store others in a place only they know. Some of the stored acorns will survive the winter and germinate in the spring thanks to their forgetful owners. Others will be buried under leaves and snow and will nourish deer and wild boars in the winter.

A little brook that trickled down from the middle of the ridge ran through the oak forest, heading for the young pine grove at the mouth of the basin. The brook then continued through a shrubbed area and into the ocean. Adjacent to the shrubbed area was a beach. Because the waves had been eating away at the land for ages, the beach was about three meters lower than the shrubbed area. Diplyak Basin is ideal for ungulates in many ways. There's plenty to eat and there's a water source, however small. There are many places to rest on the sunny southern slope.

There were tiger tracks on the beach. Bloody Mary's family had gathered. The missing female cub had also returned. But there were five, not four, sets of tracks. The fifth set of paws was very large. The width of the front paw pad was 12.9 centimeters. It was the Great King, Khajain! He had been on a walk on the beach with Bloody Mary's family. It was a truly amazing discovery. It was very rare for five tigers to gather, but it was amazing that the Great King was casually spending time with Bloody Mary's family. This had two significances: it was proof that Khajain had fathered Bloody Mary's cubs and that there was no conflict between Khajain and the male cub.

The tracks followed the brook and led to the pine grove. A few wild cherry trees stood by the brook, and a dozen young pines with thick, bushy branches grew nearby. In the pine grove, low-hanging branches made for a cozy space. The countless pine needles softened the sting of the hot sun and scattered the rays like rice grains on the forest floor. The ground was soft with layers of dry pine needles that had fallen over the years, and the low branches formed the shape of parasols, sheltering the ground from snow. It would be cool in the summer and warm in the winter here. From the pine grove, the ocean was visible over the surrounding shrubs, so it was easy to keep an eye on the area while inside the forest. What's more, there was a brook nearby. Tigers like to rest in pine forests like these.

There was a deer carcass in the forest. The flesh clinging to the bones was dry but not stiff, which meant it had been only a few days since it had died. Bloody Mary's family must have hunted it and dragged it all the way over here to eat. Since there were five of them, almost all the flesh was gone, including the scalp. We found no evidence of a scuffle here, either. The fact that there had been no conflict over food meant all five were on friendly terms.

There is one thing more threatening to a male cub than humans and other beasts—a male tiger that is not his father. Male tigers have an instinctive drive to spread their seed as far and wide as possible. They try to secure a vast territory and prevent other males from mating with the females in their domain. A tiger who has such a firm hold over his territory is the Great King.

The larger the Great King's territory, the more the young tigers born in the area are pushed to the edges of the nature reserve. Leaving the mountains spells trouble for tigers, as they run into poachers' snares and farmers' traps near the villages. It also doesn't help that they are young, inexperienced, and unable to quickly adapt to new surroundings. For young male tigers, this is the second-most dangerous phase in life after their cub years. Male tigers are far more likely to die before they secure a territory than female tigers are. If a young male tiger manages to overcome these obstacles, migrate to a new region, and establish himself as the Great King, he takes on the important task of mating with the female tigers in the region. The tiger contributes to the well-being of the species by preventing incest.

But most male tigers pushed out of their territories cannot secure new grounds, and therefore fail to spread their DNA. Even if there happens to be a vacancy, the competition is fierce. If a tiger does manage to take over a territory, he faces yet another challenge if a female in the territory already has cubs. Because female tigers don't go into heat until they've raised their cubs to adulthood, the male tiger must wait a few years to mate with the female. No male tiger, however, is kind enough to wait that long. Once a male tiger takes over a territory, he kills the cubs spawned by all other male tigers to expedite the process. A female tiger who has lost her cubs goes into heat within weeks or months.

The male leaves the moment he impregnates the female, and she raises the cubs alone. Even though they come from the same womb, male and female cubs have different relationships with their father. A male cub is a relative, but also a potential competitor. Unlike females, males are reluctant to share their territories with sons and brothers. Male cubs, therefore, frequently get into conflicts with their fathers as they grow bigger. Competition exerts a greater influence on their relationship than blood ties do. Because of this, there is a theory that if tiger fathers run into their sons before they are fully grown, they kill them. This story had spread among poachers and hunters in Ussuri, Manchuria, and North Korea.

But Khajain's attitude as he had walked along Diplyak Beach stood in stark contrast to the rumors. There were no hints of animosity in the footsteps he had left, only a firm acknowledgment that the cub was his son. The young male tiger was nearly fully grown. The family members were about six months away from going their own separate ways. The male cub was so big that its paw pad was eleven centimeters wide, large enough for the Great King to start feeling threatened. If the rumors were true, Khajain should have killed his son by now, or his son should have fled. But father and son had strolled down the beach together and shared a deer.

Dr. Yudin at the Far East Branch of the Institute of Biology and Soil Science in Vladivostok raises tigers in a confined oak forest measuring 150 square meters. He has a male tiger named Kuchir and a female named Nyurka. They've given birth to many litters together, and they're famous for being the consolation tigers that TV stations film after failing to get footage of tigers in the wild. Dr. Yudin studies tiger family relations with these animals. According to his findings, male tigers are able to stay with their families, like female tigers do. As he explained, "Male tigers also play a very important role in the cubs' education. They hunt for the cubs and watch them eat. If they're eating wrong, the male tiger demonstrates for them. No matter how hungry the male tiger is, it gives up its food for its cubs. The males, in essence, help raise the cub. Researchers were surprised by this finding."

I've observed and filmed Kuchir, Nyurka, and their cubs. There is fatherly affection in the way Kuchir teaches and looks after his cubs, but his involvement can most likely be attributed to the confined space the tigers live in. The behavior of one confined tiger doesn't represent the behavior of tigers everywhere.

As the example of Khajain revealed, male tigers in the wild sometimes do seek out family and maintain good family relationships. As long as he was with his family, Khajain would teach and look after his cubs, as Kuchir does. But this was a temporary arrangement. He wouldn't stay with the cubs and look after them like their mother. There has been no evidence of such behavior in wild tigers. The instinct to travel around their vast territories is too strong for male tigers. Khajain had an unyielding impulse to check on his territory and keep things in line, to make sure another male tiger had not invaded his territory and that the female tigers were still around, to keep track of the movement of humans, and to monitor the migration of ungulates. Perhaps, like the Gwanggaeto tigers, he was constantly searching for new lands to conquer.

There may be conflict between a father and son in the future, when the son is mentally and physically mature enough to fight over land with his father. The father might kill the son, or the son in his prime might banish his elderly father. But it was clear that Khajain was a father and not the Great King when he was on the beach with his son. He returned to his family after traveling around his territory and spent quality time with them.

Bloody Mary's family had followed the brook and headed for a ridge on the coastal range. Ussuri deer had left fresh droppings scattered in the nearby oak forest. There were a few deer and wild boar skulls and bones as well.

The tiger family had gone their separate ways at the summit of the coastal ridge. Bloody Mary had led the cubs down in the direction of Mayak Village, and Khajain had climbed the ridge inland. The East Sea undulated to the southeast, and the green sea of trees unfolded to the northwest. Khajain had set out on a pilgrimage across his land over the Sikhote-Alin Ridge. I could picture the Great King walking through the many overlapping folds of the ridge. I shivered as if a chill had passed over me, and suddenly felt alone.

CHAPTER 10

People Who Worship Tigers

After parting ways with Khajain, Bloody Mary had headed south in a straight line. If she hadn't traveled in a straight line, we might have lost her tracks halfway. Gravity pulls all creatures to the ground, but some are crafty enough to leave no tracks. Her cubs sometimes took a detour, but always returned to their mother.

We followed Bloody Mary's tracks over a few hills and discovered a lake four times the size of a soccer field. The water that flowed down the valleys of the coastal ridge gathered here and slowly made its way to the sea. This was Mayak Lake. A watery oval at the bottom of the coastal basin, it reflected the sky, clouds, and trees. The lake was a wide-open Cyclops eye of the land and a mirror of the forest, reflecting its surroundings. The sparrows that had yet to depart for their trip south glided just above the smooth surface of the lake, and a cattle egret stood among water lilies and roundleaf pondweed, occasionally reaching for something in the water with its beak. Besides that, the lakeside was quiet and deserted.

Two more hills later, we spotted a zigzagging shoreline of capes covered in oak forests. On the cape jutting farthest into the sea stood a white lighthouse. The two navy-blue buildings next to the lighthouse were the barracks of a small military contingent stationed there to protect the lighthouse and the facilities.

A thousand years ago, the soldiers here watched the ships of Korea and Japan sail by. Today, the Russian troops in Mayak do the same from the lighthouse. *Mayak* is Russian for "lighthouse." On a slope a little ways away from the lighthouse was a small settlement of ten or so households. This was Mayak Village, the southern border of Bloody Mary's territory.

There's a reason tigers do not dawdle. The border between tiger territory and human territory is a very important stop on a tiger's patrol, but it's dangerous to dwell there for long. Bloody Mary had stood on this high peak with her cubs by her side, looking down at the lighthouse barracks and Mayak Village. She had arrived at the end of her territory, observed the humans living there, and turned around.

Tiger cubs receive geography lessons as they travel with their mothers. They learn about human villages and their dangers, where ungulates frequent, and where good hideouts are scattered across valleys where they can break their journey. Bloody Mary's cubs were out there with their mother that very minute, learning something. Perhaps they were feasting on a wild boar they had hunted, or were resting. After a good rest, they would head inland toward Deer Valley or the Dragon Spine. Now that they had seen the east coast, it was time to travel inland.

We decided to conclude our expedition research at that point. We bid farewell to the tracks of Bloody Mary's family, the traces of tigers pouncing and chasing each other, and headed down to Mayak Village.

Olga Kimonko's house stood farthest up the slope. When I walked through the front door, I was greeted by Tanya, Olga's daughter. She told me that her parents were still out in the mountains.

(Most residents of Mayak are ginseng gatherers, who forage in the mountains, or fishermen.) I took off my backpack and felt so light that I thought I might fly away.

Tanya, who had a limp from a childhood injury, looked after the house while her parents were out gathering roots. She had a twin brother, but he had moved away. Tanya had finished high school three years earlier, but she seldom left the house and didn't speak much. She said she wanted to stay in the village forever and never get married.

I first met Tanya's parents in the woods. I was studying the forest when I saw an elderly Asian couple walking down the forest path toward me with a burlap sack slung over their backs. They came over and said hello as if they didn't care that they would serve a heavy sentence if they were arrested for trespassing in a nature reserve. They were Aktanka (who just went by his last name) and Olga Kimonko, Tanya's parents. Further conversation revealed that they'd been out to gather roots and that ginseng gathering was their livelihood. They said digging around on the mountain for roots was what they were accustomed to, comfortable with, and all there was to their lives. Worldly rules didn't matter to them because they were living according to the rules of nature, not the rules of man. I knew that it was illegal for them to be in the nature reserve, but I saw no reason to report them. We walked together as we talked about tigers. We instantly became close, for we shared the same views on nature and tigers. The couple invited me to dinner, offered me a bed for the night, and the next morning gave me the present of a small medicinal root before I left. Ever since then, I have been dropping by when I am in the area to exchange news and catch up.

The two ginseng gatherers returned close to dinnertime. They had gathered a great deal of roots, all of them old. It takes ginseng nearly a hundred years to fully mature, so they left the young plants and only took the fully grown ones. It was the end of August and the ginseng seed at the tip of the stigma was ripe red.

The second Aktanka and Olga saw me, they skipped any sort of greeting and jumped straight to talking about the tiger tracks they had found in the mountains. It sounded like they had found Bloody Mary and her family's tracks. They excitedly reported that they saw tiger tracks when they were in the mountains, but they'd never seen this many at once. They said they had found a lot of roots due to the auspicious tiger tracks, and thanked their god Amba.

Aktanka is Nanai and Olga is Udege, the most devoted tiger worshippers among the indigenous Ussuri. Of the Udege, she belongs to the Kimonko clan, who live on the Horr River. Aktanka had a Chinese father, but his Nanai mother belonged to the Aktanka clan. He should have inherited his father's last name, but he went with Aktanka anyway.

There's an interesting Aktanka fable: Long ago, a Nanai traveled deep into the mountains to hunt. He had to hunt through the winter, so he brought along his daughter to cook for him. They stayed in a lodge and hunted for a long time. One day, the father returned early from his hunting to see a tiger coming out of the lodge. He rushed into the lodge to find his daughter unharmed. The following year, the daughter had a son. The father named his grandson Aktanka, "born from a tiger." This boy was the progenitor of the Aktanka clan. The origin of the patrilineal line is a tiger.

Aktanka and Kimonko were worried about their son. He had moved to the city and would not return even though he struggled in poverty. He hadn't been able to adapt to city life, but he wouldn't return to be a ginseng gatherer either. The police sometimes called them about him. He was once beaten to a bloody pulp by gang members and hospitalized.

Olga had had a dream before giving birth to the twins:

> I was fishing on a river and I kept hearing birds singing. I looked around and spotted a white bird sitting on a large willow near the river. I bowed and paid my respects to the bird because it was the spirit of the willow. But when I looked up again, the bird had disappeared. Just then, the willow split in two to reveal a baby inside. I had such high hopes because the twins were willow *ado.*

Each time the couple mentioned their son they used the word *ado*, Nanai for "twin" or "twins." The indigenous Ussuri have a tradition of twin worship. They believe one in a pair of twins is a gift from the spirits. Olga believed without an inch of doubt that one of her twins was a gift from the spirit of the willow.

In Nanai myths, many heroes are born as ado. Long ago, they believe, there were three suns. All creatures died under the heat and the forest was charred black. Even fish died in the too-hot water. Finally, the spirits of the land and water took pity and sent a pair of ados.

One morning the ados went to the spot where the three suns rose, shot two down with their bows and arrows, and left one for the living creatures on land and in water. Since then, living things have been breathing fresh air and drinking cool water.

Ados were heroes in the past, but today, they're no different from other indigenous Ussuri who are trying to get by. Indigenous communities in Ussuri have been dissolved, their members scattered and burdened by a broken sense of identity. The changes that led to this crisis began in the nineteenth century.

In the turmoil of the Taiping Rebellion and the Opium Wars in the nineteenth century, the Han migrated to Manchuria in great numbers, assimilating the Manchurians to Han culture and weakening the solidarity among the indigenous tribes in the region. By the end of the century, over 80 percent of the population of Manchuria was of Han descent.

The sheer number of immigrants from China presented a problem, but more concerning was the kind of immigrants who ended up in Manchuria: illegal ginseng gatherers, miners in the up-and-coming gold mining craze, vagabond poachers, and infamous bandits. Manchuria soon became a safe haven for Chinese criminals.

When Russians began to push down from the north by invading Ussuri, the Udege and Nanai were threatened from both sides. The Russians chased them south, and the Chinese immigrants pushed them north.

The Udege and the Nanai were good targets for exploitation by the Russian merchants who came to Ussuri. The Russian merchants used the same tactics they'd perfected on the indigenous Siberians: provide them with cheap mass-produced goods and vodka in exchange for quality fur. The Udege and Nanai fell under the spell of vodka, and alcoholism ruined the lives of these hard-working, generous people. The Russian merchants' abuse of their trade relationship so thoroughly ruined the Udege and Nanai that in 1912, Nicholas II of Russia banned all fur trade activities.

But the Chinese merchants were a far worse influence: they subdued the Udege and Nanai with alcohol and opium. To get their fix, many native Ussuri with alcohol and substance dependency had to put down their next year's hunt as collateral. Within a year, the interest on their loans reached 300 percent. Chinese merchants kidnapped the wives and daughters of the native Ussuri who could not pay off their debts and sold them or made them concubines. If any debt was still left to settle, they enslaved the men. The Chinese involved in these activities were mostly bandits or criminals who imposed harsh punishments if the native Ussuri did not comply. The Ussuri who were sold in China as consorts or wives gave birth to a great number of biracial children. In China, they were referred to as *tazhi*. Rejected by both the Chinese and the Ussuri, the tazhi were never given the chance to develop a strong sense of cultural belonging. They wandered the Ussuri area without a community to keep them grounded.

But this exploitation and plunder were nothing compared to the diseases introduced by the Russian and Chinese outsiders. With no immunity resistance to fight off these new viruses, countless native Ussuri died from smallpox and other diseases. The surviving Udege and Nanai moved deeper into the forest to protect their culture and lifestyle by avoiding contact with the outside world.

The October Revolution in 1917 instigated a five-year civil war that ended with a victory for the revolutionaries and the birth of the Soviet Union. Ussuri now belonged to the socialist regime and was consequently subject to Lenin's Russification policy. During this period, the native Ussuri were forced to integrate into Soviet society and its political system.

Russification, notably, banned nomadic lifestyles and folk religions and advocated for building settlements and consolidating scattered villages. Lenin's nationalist policy, characterized by integration, collectivization, and anti-shamanism, reached its climax in the 1930s under Stalin's rule and continued into the 1980s when Mikhail Gorbachev was the general secretary of the Communist Party. Nearly all socialist agendas were pushed uniformly without regard for the traditions of the native people or the environment. These policies succeeded in dismantling the very foundation of native Ussuri communities.

The Soviet Union's first move in implementing Russification was to prohibit nomadic lifestyles, forcing the natives to settle in a fixed place. Small clans of nomads and hunters were consolidated into reservations. The Nanai who lived in a hundred different regions across Ussuri were forced into twenty reservations. Hunting clans spread all across Ussuri were forced into one village. Once the consolidation began, the main grazing fields, hunting grounds, and livestock of the Ussuri became Soviet property.

In 1922, animism, totemism, and shamanism were officially banned. It was illegal to hold shamanic rituals, carry a drum, or wear special costumes and accessories. The Soviet atheists who believed in materialism considered the nature-focused religions of the natives harmful and determined that they should be thoroughly stamped out. Government officials failed to comprehend what animist and shamanic rituals meant to the Ussuri, who were highly specialized in nomadic herding and hunting. Anti-animism and anti-shamanism policies were tantamount to ripping the souls from the Ussuri.

In the wake of these changes, serious problems began to surface within Ussuri society. Russian culture was imposed on children from an early age, starting with childcare facilities and schools. While parents were out grazing livestock or hunting, children of Ussuri learned the Russian language and culture. They were given no opportunity to learn about the Ussuri culture or way of life. Russian culture was imposed on Ussuri children from such a young age that they grew up speaking Russian as their dominant language. Starting in 1936, all men of a certain age, regardless of race, were conscripted under Stalin's constitution, and a multiracial army was formed. Most of the soldiers whom young native Ussuri men interacted with in the army were Slavic Russians, which meant a further infusion of Russian language and culture.

While it's true that the young Ussuri people were able to take advantage of Soviet-provided education, medical care, job opportunities, and other conveniences, they gained these things at the cost of their close relationship with nature. They also lost pride for their culture. The characteristic sense of community for hunters and nomads also faded and was replaced with vodka-induced alcoholism and lethargy, an increase in violence and suicide due to their loss of identity, and other societal problems. The rift between the older generation who continued to practice traditional religion and the younger generation immersed in Russian culture continued to widen.

After the fall of the Soviet Union, the Bear Festival was revived. A yearly ritual held at the beginning of the hunting season in which the blood or a part of the first hunted bear of the year is buried, the Bear Festival allows hunters to repent for taking animals' lives and pray for a good year of hunting. However, the young Nanai and Udege thought of the festival as nothing more than an opportunity for singing and dancing. Only the older generation understood the religious meaning behind the ritual. One native Ussuri writer mourned this phenomenon: "A culture that has lost its memories is a culture destined to disappear. Such indolence is the cause of a culture's decay."

It wasn't until 1985, when Mikhail Gorbachev became the general secretary of the Communist Party, that the truth about the native Ussuri came to light. Gorbachev's perestroika and glasnost eventually led to reform policies aimed at improving the lives of native populations. In 1990, the Association of the Peoples of the North of the USSR was founded (later renamed the Russian Association of Indigenous Peoples of the North), and the Supreme Soviet agreed to permanently return Ussuri's Samarga region, the main Udege homeland, to the Udege. The survival of the last animist peoples wandering the Ussuri forest was thus guaranteed. This was an important historical moment that carried great symbolic significance for the native Ussuri. The Nanai, Ulchi, Gilyak, and other native tribes soon also had parts of their forests and rivers returned to them, similar to the way American Indians were given back small pieces of their land in the form of reservations.

In this environment, once-threatening elements of native culture came to be seen in a new light. Animism gained a new status as a pantheic gesture of humility; it meant acceptance of the fact that humans cannot survive without nature and advocated that we live harmoniously with all creatures on earth. Shamans, once derided as quack doctors, became acknowledged for their understanding of nature and their wisdom to provide spiritual guidance for native members living in the modern world, and they were elevated as leaders of their communities.

The modern history of Ussuri is a story of cultural invasion brought on by misguided applications of ideology. The fate of the native Ussuri was not much different from that of American Indians and Africans, as imperialist nations divided up the land that had been occupied by indigenous peoples for thousands of years and claimed it as their own. The indigenous communities in Ussuri were first trampled upon by imperialism and then sacrificed by socialism. The Ussuri, who maintained a close tie with nature instead of ideology, were robbed of their way of life.

The old Udege still joke about the good old days when their communities were so numerous that the white tundra swans migrating from Kusun River to Olga Bay would turn black as they flew through the plumes of smoke rising from yurts. Hundreds of thousands of indigenous peoples lived in Ussuri until the mid-nineteenth century, but only ten thousand remain today. The culture that once flourished is now becoming a myth, departing this world side by side with the Ussuri tiger the people worship.

Still, the ginseng-gatherer couple continued to work in nature and protect their culture. Both descendants of tribes that worship tigers, Aktanka and Olga Kimonko had similar reverence for tigers and nature. Tanya had decided to stay in the village, unlike many other young Ussuri who leave for the cities, and maintain her close relationship with nature. She would inherit this spirituality and way of life from her parents. It was a comfort to know that the culture of the native Ussuri lived on.

PART III

To Meet, and to Say Goodbye

CHAPTER 11

Hotel Construction

IT SEEMED LIKE only yesterday that I'd heard cuckoos by day and jungle nightjars by night, yet the fall had found the Ussuri forest as swiftly as the wind. A herd of Ussuri deer in search of mates splashed across the blue river, and Manchurian wapitis on a cliff bellowed in the silver moonlight. The season of procreation had come for these animals—a time of instinct, seduction, and passing on the gene. In September, even the creatures hurrying along mountain paths stopped to listen to these cries. With the reddening leaves, the year was already in its twilight.

The time had come to choose a stakeout spot. The nut pine forest on Crow Mountain would be good, and the wild walnut forest in Deer Valley, which had also had a fruitful year, was worth considering. The forest of oaks in Santago or Dipiko would be ideal, and Urine Rock on the way up to the Dragon Spine wouldn't be bad either. The upstream region of Shauka River was probably not a great choice because we hadn't found any traces of tigers there during our summer expeditions. The northern region of the coastal ridge was not ideal because the oaks there were in their off year, but the southern region was a strong candidate thanks to the year's bumper crop of acorns.

The key was to focus on where the tigers were likely to be in the winter months rather than locations they had frequented in the spring and summer. Stefanovich and I scoped out a few places, but ultimately decided on three areas: Diplyak (warm even in the winter) Beach on the eastern coast, Azalea Cliff, and Deer Valley, located inland.

The Diplyak Basin faces south and, true to its name, really is warm even in the winter; the thick oak colony there had produced abundant acorns that year. Ungulates would appear often once it got cold, and so would Bloody Mary to hunt them. We chose two locations for the stakeout: one on the beach where the waves had carved a three-meter cliff into the wooded area over the years, and one in the oak forest in the basin.

Azalea Cliff is a treacherous rock mountain, a difficult place to build a bunker. But Bloody Mary had to pass over it to get to the coastal basin, and I really wanted to see a wild tiger walking along a cliff adorned with azaleas, like the ones I knew from the east coast of Korea. So we decided to put a bunker on Azalea Cliff.

Deer Valley, which unfurls inland from Crow Mountain, is populated with many deciduous and coniferous trees such as wild walnut, oak, nut pine, and fir, forming a superb, sun-dappled mixed forest. It's a good habitat for deer and wild boars and has long been called Deer Valley thanks to the Ussuri deer that prefer dense forests with even terrain. Bloody Mary had to go through Deer Valley to travel from the east coast to the south of the Dragon Spine. We marked out a spot for a bunker by the path upstream from Deer River, one of Bloody Mary's routes.

Now that we had chosen the areas, we had to pick the exact spots to dig the bunkers. For this, we needed to refer to the detailed eco-maps we had made. The large-scale eco-map charted the tigers' territories in their entirety and helped us identify good stakeout spots, while the small eco-maps, which contained comprehensive data on the migration routes of animals and topographical characteristics of the area, were ideal for finding optimal bunker locations. The small eco-maps contained a lot of useful information about tiger paths. Tigers approach an area through many paths, and we ranked them each according to probability of a tiger encounter. We did the same for ungulate paths as well. Sometimes, ungulate paths are harder to predict than tiger paths because, compared to tigers, ungulates travel in more erratic patterns.

We had to consider the camera angle very carefully. Cameras must be guaranteed an unobstructed view of the tiger's path and territory from the bunker. For example, the forest of young pines by the Diplyak Basin was a likely place for tigers to rest, and the cameras and bunker there had to be positioned at the same level as the path, or slightly higher, so that we could capture the surrounding trees and shrubs and watch the pine forest without having the view obstructed by the wild reeds. At the same time, the bunker had to blend in with the natural surroundings. It's not easy to find a spot that is secluded enough, yet also has a clear view. Our final consideration was about environmental impact; the area had to be restored to its original condition after the stakeout period to minimize damage on the natural surroundings.

We started building the bunkers in the coastal areas. It generally takes four to five days to build a bunker, so we camped out in tents for a few days during the construction period. We were anxious that the tigers might find out what we were up to, so we tried to get the job done as quickly and efficiently as possible. We kept noise down to a minimum when we were digging or talking to each other, and we wrapped things up by four in the afternoon when tigers become more active. If the tigers were to hear us, hide themselves in a nearby bush, and watch what we were doing, they would change their migration routes and all our work would be a waste.

When we build bunkers, we also stay aware of the tigers' whereabouts, however approximate. Bloody Mary was inland with her cubs at the moment, so it was the perfect time for us to build bunkers in the coastal regions.

We called our underground bunkers "hideouts" or "hotels." Naming them something pleasant helped ease the discomfort of our extended confinement. We tried to make the inside of our bunkers as comfortable as the inside of hotel rooms. To survive the long stakeout period of over six months, we arranged things efficiently in the cramped spaces and insulated them meticulously to keep out the northwesterly winds, which could drop to as low as -30°C to -40°C in the winter.

Most of our hotels measured 2 by 2 meters with a 1.8-meter-high ceiling. We had to stoop a little when standing up, and our heads and toes touched the walls when we lay down. If the conditions didn't allow for a bunker that size, we sometimes made them half as small. We called these little bunkers "motels." They were so cramped that whoever was assigned to them had to curl up in a ball to sleep every night for six months. It would be too taxing to endure six months of winter in a space smaller than that.

Even though the bunkers were small, their construction required more building material than one would think. We transported our basic equipment and materials from our base camp to the bunker site, but most of the building supplies came from the surrounding area. We couldn't have transported everything to the forbidding terrains where the bunkers were built, since we would have had to carry those supplies through many parts of the journey. There were already useful building supplies in the forest or on the beach. We found fallen tree branches that hadn't rotted through yet and collected ship masts or pine panels washed up on the shore.

Once we had our materials ready, we dug a trench of 2 by 2 by 1.8 meters and put poles in the four corners. We usually built the bunkers on slopes, so the 1.8-meter poles went all the way into the ground in the back, but only 1.4 meters deep in the front. This was where we built a 50-by-40-centimeter entrance. We then made a roof from branches and pine boards and reinforced the inner walls with wooden panels so they wouldn't cave in. We installed a raised bed 60 centimeters above the floor in the back half of the space.

Now that the basic structure was ready, it was time to work on the interior. First, we wallpapered the rectangular frame of tree trunks and pine panels with blankets and cardboard boxes that didn't smell of chemicals. We then installed shelves on the left, right, and back walls. The shelves on either side were for food, cooking utensils, and other things we would use every day, and the back shelves were for books or equipment that wouldn't be necessary right away. The space under the bed was chosen as a storage space for fuel and an excrement container. But not everything could be stored in the open. Batteries quickly lose their charge and drinking water tends to freeze in cold weather, so we dug holes on either side of the bunker, insulated them with plastic foam, and stored these items there. Filming equipment is vulnerable to humidity, so we strung up about ten bags of silica gel from the ceiling, making the space look like a shaman's hut.

I placed the tripod near the entrance, attached the camera, and sat the way I would if I were filming. I built another shelf in easy reach from that position and set the filming equipment there. That way, if a tiger appeared, I could easily grab a spare battery or tape. Since I couldn't have the light on in pitch darkness, I had to keep things within reach or I might drop something while fumbling around and give myself away to the tigers. It was important to remember where things were and to always put them back in the right place. Things that might roll off the shelf were hung on a nail in the wall. I put seven nails in the wall, numbered them one through seven, and memorized what was hung where. I also put myself through darkness training by grabbing things with my eyes closed. If I wasn't vigilant about these minor things, it could lead to bigger problems later.

We draped three layers of thick blankets over the entrance and made a thirty-centimeter hole in the horizontal middle of each for the camera lens. The holes in the first and third blankets were also dead center vertically, while the hole in the second was offset ten centimeters lower. That way, when the camera lens was in place, the blankets kept the air out so the inside of the bunker remained well insulated and not as exposed. The earth-tone colors and leopard print of the blankets blended in with the natural surroundings. Pine panels were nailed together to make a door that fit the opening perfectly. The door doubled as a dining table by day.

While one team was on stakeout, the other team remained at base camp and delivered supplies to the stakeout posts and studied animal tracks. These tasks were also important. We used the lodge in Petrova as a provisional base camp in addition to the one in Kievka, and the Ural base camp was in constant operation. We alerted each other whenever something happened. Beside each bunker, we installed a small antenna camouflaged as a branch so that we could communicate between the stakeout bunkers and the base camp. Since the camera in the bunker could see only what was in front of it, we also installed three small cameras aimed at the sides and rear of each bunker.

Our last step was to camouflage the exterior of the bunkers. It was just as important to make the bunkers inconspicuous from the outside as it was to stop scents, sounds, and light from escaping from the inside. We planted bushes on both sides of the entrances and insulated the roofs and side panels with several layers of plastic before burying each bunker under a generous layer of dirt. It was warmer that way. We sprinkled the dirt layer with leaves and added an old fallen log on top; with that, the bunker camouflage was complete. It was good enough to fool the animals that frequented the area.

It took all of September to build the four bunkers where we would spend two winters over two years staking out in those locations. Nervous that the tigers would pay us a visit while we were working on the bunkers, I felt relieved once the preparations were done. No amount of camouflage could completely erase the footprints or scents of humans. It would take the first snow to finally wipe all scents and tracks clean. By then, each hotel would be part of a snowy landscape, and the insides would be as cozy as a bear's den.

CHAPTER 12

Cold Weather, Green Pines

An eagle perched on top of a tree. The feathers on its back fluttered in the wind. In the fall, the fish swim against the current, but the birds turn their backs to the wind. A badger crossed a log bridge over the brook to dig through leaves for food, and an Asian black bear climbed nut pines to gather pine nuts. Animals preparing for hibernation busied themselves in the autumn forest, trying to accumulate that extra ounce of fat for the winter. In a single leaf falling on a windless day, time folded its long stretch into a moment and slipped past right before my eyes.

It was early October, and the stakeout phase had begun. It was cozy here in the two-square-meter underground bunker in Diplyak Basin. The beginning is always exciting. Sitting still and watching nature sounds very romantic. But after fifteen years of staking out in underground bunkers, I felt that "routine" had replaced "romantic" as the more fitting adjective.

While many researchers study the flora and fauna of the forests and record the traces animals leave, that is only the first half of my research. The second half is to observe and record the lives that leave those traces. The second phase, the stakeout, is the main part of my work.

You can see and experience a good deal of nature simply by being in the mountains. You can watch blue-and-white flycatchers or squirrels. You can feel the wind in your hair and savor the experience of being in the forest. These little joys are uplifting in themselves. But if you want to access the intimate depths of nature, you must become a tree on a slope. If you wait patiently and are as still and quiet as a tree, nature will show its soft underbelly. What daily headaches do chipmunks and blue-and-white flycatchers deal with? How do trees and the forest change clothes with the seasons? How does the wind turn from friend to foe, and when does the fog come for a visit and how long does it stay? Only when you become a part of nature does nature reveal the answers to these personal questions.

Tigers leave traces of their activities in the forest that teach us a great deal about them. But that is not the same as seeing a tiger. Unless you're very lucky, the tiger always senses your presence and leaves before you have the chance to catch up with it. To see a tiger, you must stay in one spot. Instead of chasing after it, you must wait for it to come to you. The tiger sees you when you're traveling around on expedition research, but you see the tiger when you're hidden in one spot on stakeout research. During expeditions, I become an agent, but during stakeouts, I am one with the subject. As an agent, I can cover a great area but miss some details. As a subject, I cannot cover a great area, but I see the details up close.

Staking out in nature requires endurance and self-restraint. You must take the time to acclimate to the nature around you or you'll feel restless and claustrophobic. Staking out involves being confined to a two-square-meter underground cell in -30°C weather for six months, unable to take a shower, shout, or turn the light on. Willingly taking on the challenge is the first step to becoming one with nature.

You must have faith. Walking through the woods, you often come across owl pellets, small masses of indigestible bones and feathers that have been regurgitated by the bird at the foot of a tree. When you find one of these, you know an owl is sitting on a branch over your head, looking down at you. You may be overcome by the urge to look up and see the owl for yourself. But the moment you give in and look up, the owl will fly away. I trust the owl is up there and continue on my way. This way, the forest avoids a small disturbance and maintains its peace. Trusting an animal is there by looking at its traces rather than pursuing the animal itself: this is faith in nature.

Once you begin to lose faith, a spark of doubt appears. If an animal doesn't show up when it's supposed to, you become anxious. As you explore every possible scenario, ranging from the reasonable to the extraordinary, your common sense comes undone. Your objectivity dissolves and fuels the doubt with disquiet and irritation. This happens in an instant. And that's when the researcher bursts out from the stakeout bunker.

TWENTY-ONE DAYS IN, and snow fell relentlessly, like feathers from the sky.

I heard the slushy sound of snow melting in the ocean. The beach was buried under a white blanket. Waves rushed in and licked at the edges. Thick piles of snow grew on every coniferous tree branch, plopped to the ground, and grew again.

The wind gained momentum and sent the falling snowflakes whirling through the air. Each year at the end of October, it's as if the door to a massive ice cave opens and cold winds rush out from within. The Ussuri refer to this northwesterly wind as "white hair wind," and I imagine a witch with white unkempt hair who flies four thousand kilometers to the east and south, driving the snow. The rivers freeze solid, the forest is buried in snow, and the six months of winter begin.

The northwesterly wind from Siberia meets with the air currents coming off of the East Sea to decide the Ussuri winter weather. Stronger northwesterly winds bring a cold winter, while a stronger easterly wind brings a mild one. If the two collide, it snows. The first snow is usually just a dusting, but this winter was starting off with a snowstorm.

It was the first day of November. The moon was bright. It must have been the full moon, or maybe just after; the moon looked a little dented. Under its light, the snowfield was as bright as a field of fireflies. The moonbeam struck the thicket from an angle, bathing one side with light and casting a shadow on the other. A clean line was drawn between light and dark. Each time the wind blew, snowflakes on the bushes became airborne and then fell on their own shadows.

In the bushes, orbs of light danced in midair. Moments later, small furry raccoons rolled out onto the forest path and blinked their shiny eyes. They moved along haltingly, looking out for danger. Their round shadows waddled along behind them on the white snow. Having traded their summer coats for thick winter ones, they all looked fluffy. The brave father led his two chubby children, followed by their mother, whose thick rings around her eyes looked like sunglasses. She doddered along, watching their backs. Their den must have been nearby. I often saw them scampering down this forest path to hunt for food an hour or two after sundown. "Hunting" isn't the right word to describe what they did: they rooted through the forest all night, eating everything they could get their paws on, and returned home at dawn.

As the raccoon family disappeared from view, leaving behind an echo of paws crunching against snow, the Eurasian eagle owl began to hoot. The sometimes whining, sometimes beguiling messages exchanged between owls filled the empty basin: *Have you had dinner? I love you still.* An owl occasionally spread its big, light-soaked wings and cast a dark shadow on the ground as it flew soundlessly over the basin. If an owl hooted in the south, another would hoot in the north. Two hoots in the south, two hoots in the north. Then one owl flew over to the other and went *whoo whoo whoo*. The other owl responded with a nasal *aaaah*. The Eurasian eagle owls were mating. Their coitus was brief, and after it was over, the basin fell suddenly into silence. I tensed up and peered into the viewfinder, wondering if a tiger had appeared. It was quiet all around. The only movement was the wind, which blew through the basin and shook the bushes.

I was just learning to tell different waves apart by their sound. Today's pattern: three or four sets of six small waves followed by a big one, and then one set of five small waves followed by a big one. Repeat. Listening to the sound of the same waves at the same spot for over a month, I could hear the difference between the small and large ones. This was a sign that I was adapting to stakeout life.

When I first came to live in the bunker, I enjoyed listening to the waves. I felt like I was on vacation. But two weeks later, I started to get sick of them. I thought, *Can't someone turn off that monotonous droning?* I became irritable. But at a certain point, I stopped hearing them. The waves were always rumbling and crashing in my ear, so I drowned them out. Now I found myself unconsciously counting waves. I would lie wide awake in the cold, silent bunker and count, *One, two, three*...

During the day, I spent my time deciphering different winds by their sound. Sometimes I heard what sounded like a raging torrent, and other times whizzing arrows, the sound of a baby breathing, or the ocean rumbling.

I also tried to guess how strong the wind was by looking at snowflakes through the viewfinder. When the snowflakes fell in a straight vertical line, the wind was at zero on the Beaufort scale. At one, the snowflakes fluttered. At two, they fell diagonally. At four, the wind swept the snowflakes off the ground, and at seven, the snowflakes flew horizontally. The world looked like the scene inside a snow globe when the wind was at eight.

I ran out of things to read, so I read the label on a can of green tea:

> INGREDIENTS: 50% green tea leaves, 50% organic brown rice (grown in Korea)
> NET WEIGHT: 37.5g (1.5g x 25 tea bags)
> FOOD TYPE: Tea
> PACKAGING: Filter paper—100% natural pulp, box—SC Manila
> PRICE: 2,500 won
> Store in a cool, dry place for best flavor.

I never used to pay attention to green tea labels because the world is full of more interesting things to read, but in the bunker, this was entertainment.

I read and re-read the books I had brought. The quality of the book didn't matter. What was important was that there was reading material at all. Any book you read many times over will seem like great literature. You start to see the thought and care that went into its creation. I was grateful to all writers who ever wrote.

I boiled water in a small kettle and melted a ball of rice that had frozen solid. I had seaweed, dried fruit, and jerky with the rice. Stinky food was prohibited. We would never be able to hide the smell from the sharp noses of tigers. Out in the world we may be picky eaters, but in the bunker, we were just grateful that we had enough food—just as we were happy to have anything to read.

Life in the bunker helped sort out what was important and what was not. What the world considered important seemed trivial in here, and what was trivial out there was revealed to be truly important here. In a bunker, you learn to differentiate between the essential and the superfluous, the meat and the shell, the real and the fantastical.

When people are ailing or inches away from death, they turn to religion. They can see what was or wasn't important in life. Something similar happens on stakeout: in the bunker, you can feel the end of your life coming toward you one step at a time like the firm footsteps of the Ussuri tiger. It's an experience that encourages you to do the things you want to do before your life is over.

Caught in the web of life, it's hard to separate the details from the crucial points. Being cut off from the world gives you the space to reflect and remember, just as one notices the green of the pines only after it has grown cold.

CHAPTER 13

The Challenge and Response of the Ussuri Forest

At some point, a few mice began frequenting the hotel. There were four of them one day, and one was a bit large. They appeared to be a family of three siblings and their mother. They were striped field mice, golden brown with black stripes running down their backs. Unlike most rodents, field mice don't store food for the winter and must instead resort to searching for food in the freezing cold. The warmth of the bunker and the smell of food must have lured them in from the frozen ground—or perhaps I was the one invading striped field mouse territory.

I started to observe and record the mice's behavior. The largest of the four bored a hole in the wall and peeked out. The others were busy ripping bits off the blankets we had used to painstakingly line the bunker walls and were carrying them back to their nest. The large one climbed down the wall. It sniffed and examined the back shelf where the books and video cables were. Mice are very fond of chewing on video cables. I suppose it's nice to have some soft plastic after gnawing on hard wood.

The large mouse crawled along the wall, hopped onto the side shelf, and beelined for the spot where the green tea, dried seaweed, and meat jerkies were stored. She had a keen sense of smell. The lids were shut, so she started to chew her way into the containers. I reached out and offered her a bit of rice ball. She darted back about a meter, then turned to look at it. After assessing the situation, she returned to eat the rice ball. The others hesitated for a moment before they came crawling up as well. They were brazen.

The mice already knew where everything was located inside the bunker. Instead of wandering around in search of what they needed, they went straight to the spot where they knew they would find what they were looking for. When they were hungry, they went to the side shelf where the food was. When they were thirsty, they crawled under the bed where the water was stored, and when they wanted to grind their teeth, they climbed onto the back shelf where the video cables were. People think of apes and dolphins when they think of the most intelligent animals, but I think rodents might be pretty smart, too.

The striped field mice began to play, but froze when they noticed me looking at them. They observed my pupils, movement, and energy and tried to predict my next move. Was I going to get up? If I stayed still, they resumed their fun. They ran around the bunker and dove off the shelves onto the soft goose-down sleeping bag. They were the bunker comedy routine. Entertainment. A source of comfort through the tedium of the stakeout.

Yet the mice were getting too bold. They were fearless. I once did a similar stakeout underground in Korea, and I had found the striped field mice there behaved very differently from the ones in Ussuri. The Korean mice lived in a field near a village and were sensitive and wary. They appeared only when there were no signs of people, and once chased, they didn't show themselves again for a few hours or even for a few days. Their long history of encounters with people and the harm they did conditioned the mice to be cautious.

The Ussuri mice, on the other hand, must have had no such negative experiences with people. The bunker was in the wild, with only two or three villages within a fifty-kilometer radius. Wild animals rarely come across people, let alone those who sit underground in the dead of winter, so these mice didn't have prior knowledge of how to behave under such circumstances. This was why they weren't afraid of people. I was reminded of a Korean saying: "A newborn puppy knows no fear of a tiger."

The mice also made me think of a fox I had seen in the Kuril Islands. The arctic blue fox lives on an uninhabited island named Ushishir, located in the middle of the Kurils. The fur of this fox is white in the winter; in the summer, it is brown in the sunlight and gray-blue in the shade. It is a crossbreed of the arctic fox and the red fox that the Japanese army brought to Ushishir during World War II to produce fur for military use. The Japanese army retreated after the war, and the foxes have thrived for over sixty years in the wild without human interference.

The arctic blue foxes had been less guarded than the Ussuri mice. When I approached, they came over to me to check me out instead of running away. When I explored their lair to film the kits, the fox mother regarded me with as much interest as a cow would give a chicken. One of the kits suckled on my finger. Being at the top of the food chain on an island uninhabited by humans, they had no reason to fear us. If humans were to encounter highly intelligent aliens, wouldn't they behave like arctic blue foxes, too?

SIBERIAN TIGERS TODAY are extremely cautious. They loathe interaction with humans and avoid all manmade structures and objects. Past records show that they haven't always been this vigilant and circumspect. I wonder how much suffering the mighty tiger has been put through to become this way. Fortunately, despite years of hardship, some tigers survived. The ones that lived had ample experience with human threats and were in turn highly cautious. These tigers passed down their knowledge to their offspring along with the genes that allowed them to survive human attacks.

When railroads and roads were first built in Siberian tiger territories, many tigers that had never seen trains or cars before were killed in accidents. The tigers that survived taught their cubs how to cross roads and railroads. *When a large thing that smells like metal is coming toward you, don't cross. Cross at night when there are fewer large metal things coming and going. Even when you cross at night, cross only when there's no light.* Thanks to generations of education and good genes, now there are next to no tiger roadkills unless the tiger is a very young cub or disabled, and Siberian tigers cross roads or railroads only at night when there are no lights.

Tropical tigers are different. For them, the weather is mild and food is plenty, so they don't need the expansive territories that Siberian tigers occupy. While a female Siberian tiger's territory is on average 450 square kilometers, a female Bengal tiger has a territory of only about 20 square kilometers. This exposes Bengal tigers to more frequent encounters with humans, and they are thus less wary of people. One can easily observe wild tigers in India from a tour bus. Tigers saunter up and down manmade roads and nap in the shade on the side of the road. When people approach them in jeeps, they sometimes come over to sniff the vehicles and urinate on them.

Bengal tigers have also suffered quite a bit at the hands of humans, but it has made them more aggressive rather than cautious. This may be the result of the tropical climate and the close proximity of tiger and human territories. When facing threats from humans, Bengal tigers prefer confrontation to ambush. And when a tiger is injured and unable to hunt, it becomes even more violent toward humans. The Champawat Tiger became famous for attacking 436 people up and down the India–Nepal border in the late eighteenth and early nineteenth centuries. A similar tiger appears in Rudyard Kipling's *The Jungle Book*—a crippled tiger named Shere Khan who tries to eat Mowgli at every opportunity. Perhaps a real crippled tiger had been a model for the fictional one.

A hundred years ago, guns were introduced as a revolutionary way to hunt tigers. The tigers were initially as unimpressed with guns as they had been with spears and arrows. They didn't run very far at the sight of hunters, and when pursued, they fought back as they had before. Countless tigers died, and tigers slowly became alert to the dangers of guns. They became conditioned to fear the sound of gunshots and the smell of gunpowder.

Except in extraordinary circumstances, the typical Siberian tiger today flees at the sound of a gun. It either runs far away or circles back to the poacher and attacks from behind. Tigers naturally came to favor ambush over confrontation. They learned to move more stealthily and to have fewer encounters with people. Poachers these days don't get to see a single tiger, no matter how much they roam the woods with their high-performance magazine rifles. There are certainly fewer tigers than before, but the tigers have also adapted to their new circumstances.

In response, the poachers began to install gun traps and special snares. Many Siberian tigers, most of them young with relatively little experience, were sacrificed to these contraptions. But the tigers found a way to avoid the traps as well. They learned to avoid or destroy all objects in the woods that smell like metal. When destroying a camouflaged gun trap, the tiger goes to the back of the gun, fires it by hitting the stock, and then breaks the gun. They learned this from the accumulated experience of being shot while fiddling with guns from the barrel end.

To film tigers and ungulates, I have been installing cameras in the Ussuri forest for nearly ten years. The tigers found twenty-three of the cameras I'd fastidiously hidden, and destroyed them.

I rubbed deer droppings on the small cameras, hoping to mask the metal scent, but they still managed to find and destroy nearly all of them. The act of destruction was never caught on tape, only the sound, because all the cameras were attacked from behind. The dark lens of the camera must have looked like a gun barrel to the tigers. The more experienced tigers were more adept at finding the cameras than the young ones, but all tigers, newly independent or mature, dismantled the cameras they found from the back. This is proof that mothers educate their young about guns from early on.

We also found that male tigers were more likely to destroy cameras than female tigers—males tended to break cameras, whereas females preferred to avoid them. For the females that did destroy cameras, two out of three were accompanied by their cubs. While it's hard to confirm on such a small sample size, it seems females traveling with their cubs tend to be more sensitive to danger and prefer to preemptively remove things that may pose a threat to their offspring.

Tigers also destroyed sensors installed to trigger hidden cameras. There were about as many broken sensors as broken cameras, but a slightly different group of tigers were involved. Unlike cameras, sensors are made of plastic and therefore don't smell like metal, and they're less threatening to tigers because they don't have dark lenses that resemble gun barrels. This may have been why a larger percentage of females and young male tigers broke the sensors.

The safest bet with strange objects in the forest is to do as female tigers do and avoid them. Young tigers don't have the experience to deal with strange objects when they encounter them, so they are most vulnerable. Safety awareness is most acute in mature females, followed by mature males, young females, and young males.

Poachers are always coming up with new ways to hunt tigers. With the introduction of each new method, there are always tiger casualties, especially among cubs and young males. It was the same when the special snares and gun traps were first installed. It's not easy for tigers to survive the period it takes to adapt. But the tigers eventually learn. Through their evolving knowledge, they find ways to protect their species against human attacks.

The challenge and response of the Ussuri forest—the cutthroat fight for survival between man and tiger—continues.

CHAPTER 14

White Moon, White Snow, White Sky

THE SEA HAD been raging for a few days. Waves as big as houses crashed on the shore and roared as if they were prepared to fight to the death. Snowflakes whirled in the air. Caught in the strong winds, the heavy snow fell horizontally. The cold wind from Siberia met the humid air current of the East Sea to generate a snowstorm. Visibility was as low as a hundred meters. I filmed the snowy scenery and then crawled into my sleeping bag to curl up like a hibernating bear.

The number of striped field mice had increased to seven or eight. They continued to grow ever bolder. One scaled my back and climbed to the top of my head, and two knocked a cup off the shelf while trying to crawl into it. I got up and clapped my hands hard, and they fell still instantly. I picked up the cup, wiped it clean, and put it back on the shelf. Bit by bit, I began to hear their little scuttles again. They were starting to become a problem. I'd stopped doling out snacks a few days earlier, but their numbers were still increasing too fast. It was a conundrum.

Sometimes, the mice helped. I couldn't sleep too deeply, because I never knew when a tiger would show up. I needed to be able to notice any subtle change in sound or movement and wake myself up instinctively. After midnight, however, I couldn't help but fall deeper and deeper into sleep—until the mice came along and woke me up, that is. They pushed a roll of toilet paper off the shelf or fought over teaspoons. If that didn't do the trick, they crawled all over my face.

The mice also helped when I felt my heart was about to explode from unspeakable frustration. It calmed me to just watch the mouse circus romp about comically. The sounds of wind and waves assuaged some kinds of gloom, but the comedy show these little mammals put on also helped lift my spirits and distract me.

The snow stopped in the afternoon. The wind calmed and the waves also regained their composure. The clouds rushed eastward. It was the backblast after the storm. Once the backblast drove all the clouds away, the cold Siberian Anticyclone would take its place and keep the weather clear for some time. During the storm, a large ship had dropped anchor about two kilometers away from the beach, and it hadn't moved since the day before. I had been living in the bunker for two months, and this was the first time I had seen something like this.

The next morning, I found a long trail of pugmarks on the beach. A tiger had passed by in the night while I was asleep. The tiger must have preferred the sand over the thirty centimeters of snow, for the prints were right at the edge of the beach where the waves had washed the snow away. From the looks of the clean pugmarks untouched by waves, the tiger had likely been there early in the morning after the tide had receded.

I zoomed in on the pugmarks. I could see them better then, but I couldn't tell their exact size through the viewfinder. They looked like they belonged to a male tiger, but I couldn't be sure. I wanted to go down to the beach and examine them, but I restrained myself. I couldn't set foot outside the bunker. The tiger might have still been around.

I slowly turned the lens to look at the forest of young pines. Every branch was adorned with snow. There was a thinner layer of snow under the pine branches. Occasionally, snow slid off the branches and fell onto the ground, where it piled up into gentle hills. Apart from the trace of snow, the pine grove was undisturbed.

I gave up trying to guess the identity of the tiger by looking at its pugmarks. I was lucky the tiger hadn't sniffed out the bunker. Siberian tigers are generally nocturnal. They patrol their territories and hunt by night, then nap for several hours during the day on a sunny mountain ridge with a clear view of their surroundings. Up there, they track the migratory patterns of deer and wild boar and watch people's activities. The tiger that had left those pugmarks that morning might have been sitting somewhere on a ridge above the coastal basin, looking down at the coast.

The valley plunged deep toward the end of the mountain ridge that extended far beyond the basin. The wind echoed as it reverberated down the valley, carrying the scent of snow. The northern ridge was the path the northwesterly wind traveled, rousing yesterday's snow like a dragon spiraling up to the sky each time the wind hit it. Beyond the whirlwinds, bright rays of sunlight glittered as they streamed down through the clouds. Moisture froze instantly and hung in the air. This is known as the diamond dust effect. Snowflakes and diamond dust flew in the air, sparkled like light through a prism, and slowly landed over the oak forest.

To maintain a half-awake state at night, I must sleep during the day. I've done stakeouts for many years, but it's challenging every time. This stakeout was almost at the two-month mark, but my mind and muscles had not yet completely adjusted to a stationary lifestyle. The active energy of the expedition period needed to be reined back to a calm, sedentary energy. But I remained curious about the owner of those pugmarks.

WHOO... WHOO...

Gwuhh... gwuhh...

A pair of eagle owls made a nest between the boulders somewhere on top of the mountain directly facing the bunker. I couldn't hear their cries for a few days due to bad weather, but one night I heard the female asking the male for food. They must have started brooding after the recent mating period. Their exchange from when they first met until the female laid eggs sounded like *whoo whoo*. Now that they had mated and the eggs had been laid, the female's *gwuhh gwuhh* sounded a bit like nagging. She was brooding and demanding food.

The temperature dropped to -24°C. The days grew darker. I swapped out the day lens for the night lens. As usual, I kept my eye on the forest and beach through the viewfinder.

A family of raccoons walked by. The mother trudged on, followed by two scuttling cubs. The father wasn't around that day. The mother was doing a great job, as the fluffy, chubby cubs looked lovely each time I saw them.

Feeee... feeee...

The mother raccoon, her voice too frail for her figure, was hurrying her cubs along when she suddenly stopped in her tracks and looked to her left. Her cubs were behind her, but she kept her eyes to the left. Suddenly, she darted for the dry bushes to her right. The cubs waddled after her. I slowly turned the lens to the left. About two hundred paces away, in the shadows of the forest of young pines, a large creature stood still. Behind the creature, something vine-like was waving. A chill crept over me. The large creature stepped out in great strides. It was a tiger. I calmed my quickening breath. I opened my eyes wide and looked again. It *was* a tiger. At last, a tiger had come. This unchanging fact sped the breath I managed to calm. I looked at the clock on the camera. It was November 27, 19:47.

I slowly zoomed in, trying to steady my nerves. I saw its face. There was tension in its eyes. The thick fur around its neck indicated it was a male. Its shoulder blades were striking, but not large enough for a mature tiger. A young male! It was Bloody Mary's son. He had returned to the Diplyak coast for the first time in three months. He was much larger than a fully grown female now. He also had the personality of a male tiger. He sniffed around the second he emerged from the forest of young pines, then headed around to the back of a small sensor hidden by the path, and broke it. He kept examining the area to see if there was anything unusual. I didn't think he was particularly interested in where I was hiding, but he glanced in the direction of the bunker every now and then. The small plastic sensor that hadn't been there before, the faint but foreign smells that permeated here and there—these little things combined gave the place a different feel than he remembered.

The moon was hiding behind the clouds, blurring the image of the tiger through the night-vision UV camera. The camera now showed 19:51. The clouds suddenly parted and moonlight poured down. The full moon shone on the white snow and made everything clear. It was beautiful. Before me, the abyss of the East Sea glittered silver as it reflected the full moon, and all around, the snowy ridge embraced the coastal basin. The cries of eagle owls, the hallmark of nocturnal solitude, echoed from the frosty ridges and reverberated out to the far reaches of the East Sea.

White moon. White snow. White sky, I thought. White moon. White snow. White heaven and earth. And in the midst of it all stood a tiger. Bloody Mary's proud son was bathed in moonlight. The frosty, white beauty gave me goosebumps.

After the full moon emerged, the tiger turned and disappeared into the pine grove. The last I saw of him was his long tail knocking snow off a bush—19:53. I felt empty. It was as if a red tiger that had graced the middle of a traditional Korean watercolor of a snowy landscape had suddenly vanished. The moon sailed to the top of the sky, making the coastal basin increasingly brighter.

ONE IN THE morning. I opened my eyes at the sound of something treading on snow. The sound was coming from right near the bunker.

Crunch. Crunch.

The sound was slow and deliberate, as if someone was pressing a large rubber ball into the snow. I held my breath and gently turned on the rear-view camera. A bleak light radiated from the small monitor; it stung my eyes. The screen came into view as my pupils contracted, and the camera displayed a soft silhouette. It was a tiger. Bloody Mary's son was back. I saw other silhouettes behind him—two smaller female tigers. Their bodies were firm and fully grown. The beautiful tigers Bloody Mary had raised were now all grown up.

The three siblings walked toward the bunker. The male led and the two females followed. I was on edge. My heart raced as they came closer. Had they picked up on my scent? I stared into the dark screen.

But the male's gaze was fixed on the sea, not the bunker. It was the ship! The tiger was watching it float offshore, waiting out the storm. I let out a sigh of relief. The three siblings stared, transfixed by the light coming from the vessel as they quietly walked toward the beach. When the male stopped, so did the females. They were now mere meters away from the bunker. Standing calmly, they kept their eyes on the water, their intense gazes absorbed by the silence of the night sea.

Hrrrdt... Hhhhh...

One female made a whinnying sound, like a wild horse, and rubbed her face against the male's neck. As if on cue, they all began to snort and rub their faces against each other's necks. This behavior is a sign of affection among tiger family members.

The male took a few steps toward me. *Plop!* I felt a great weight pressing down on the roof of the bunker. He was lying on the flat, snowy patch right above me. The pine roof bounced from the shock. My heart jumped as well. The females also lay down on the snow and resumed gazing out at the sea.

The light from the ship docked far away flickered on the undulating waves. Mixed into the rhythmic beating of the waves was the sound of the tiger breathing, right above my head. It sounded like the winter wind whipping the branches, and I could hardly breathe myself.

Whooonnn...

One female yawned like a housecat, and the other female followed. The male also yawned and rolled over. The roof bounced again. Completely unaware that there was a human under the ground below him, he enjoyed the night, rolling this way and that. The roof moved a little each time. It was only thirty centimeters thick, including the layer of dirt over the pine panels. Trying to be as quiet as the dead only made me breathe harder. Tiger and man breathed together, the tiger aboveground and the man below, with only thirty centimeters between them. We sat on a snowy seaside hill underneath the moonlight listening to the waves of the East Sea. A beast of the wild and a man from the secular world, sharing a moment—it seemed wrong. I felt apologetic toward Bloody Mary's son for deceiving him.

About ten minutes passed. Once their surroundings fell still, the mice came out to play again. They scurried about in search of food. I was nervous.

Please stop moving, I pleaded inside my head. But the mice didn't care, and they ran about, squeaking.

Please, mice...

That moment, a mouse knocked a container off the shelf. Bloody Mary's son jumped to his feet. A chill shot down my back. He started to walk. The roof moved again. The mice instantly fell silent.

Crunch. Crunch.

I heard paws on snow. He was circling the bunker, sniffing the ground. He was just around the corner from the entrance, but he wouldn't be able to get in through the narrow opening. I used this fact to comfort myself and pressed my hand against the wooden panel blocking the entrance behind the camouflage blankets. I used my free hand to grab the rope attached to the gun loaded with blanks.

Crunch. Crunch.

He came closer, one step after the other, and stopped. He was standing at the entrance. Sound. Light. Breathing. Everything was still. But my heart beat faster and faster.

I heard him exhaling sharply through his nose. My hairs stood on end, and my fingers involuntarily clenched the rope.

Tap. Tap.

He knocked on the door with his snout. The touch of the tiger's snout on the other side of the panel permeated through and coiled up the backs of my hands like a serpent. It took all the willpower I could muster to stop myself from firing the blank. If I fired the blank, everything we had worked for would have been for nothing. Once the tiger figured out that humans hid under layers of dirt that smelled like this, the stakeout here and all the bunkers we had built inland and along the shore would go to waste. If they found an area that smelled or looked even remotely like this bunker, they would cross it off their list and find a different migration route. Besides, there was no guarantee that firing the blank would save me. Most tigers run away when a blank is fired, but each tiger is different. Some attack, consumed with aggression.

I let go of the rope, afraid I would pull it by accident. Still suspicious, the tiger continued to sniff at the entrance. The mice began to stir again. I was dying inside. Under such dire circumstances, the lives of the mice I'd done a decent job of pretending to respect seemed very expendable.

Hrrrdt hrrrdt... hhhhh...

A female came over to the male and rubbed her neck against his. The male stopped sniffing the entrance and rubbed back, snorting. The male let his guard down. He walked off, and the others followed. My hand cramped. My chest tightened. Strength drained from every part of my body as I let out a long sigh. Not enough experience. I didn't have enough experience, and neither did the tigers. Bloody Mary's children weren't veterans yet. Things would have been different if Bloody Mary had been there.

Graawr! Graawr!

The male roared as he walked down the snowy field bathed with moonlight.

Grh-waa! Grh-waa!

The reply came from the top of the mountain across the way. Bloody Mary! I quickly hit the record button and listened closely.

Graawr! Graawr!

Grh-waa! Grh-waa!

Perhaps I was biased because I was aware of their relationship, but Bloody Mary's cries struck me as more insistent, like a worried mother telling her children to come home, whereas her children's cries sounded more passive and petulant, like children dragging their feet and grumbling about being called in for supper. The roars they exchanged carried far out to the East Sea over the sound of the majestic waves. The cries of the cute eagle owl couple that had filled the basin with life until evening could no longer be heard, and the sounds of paws crunching on snow and the tigers' roaring and whining slowly grew distant. It was the night of Bloody Mary's family. One son, two daughters. I named them: Wolbaek for the youngest, Seolbaek for her older sister, and Cheonjibaek for the firstborn son. "White Moon." "White Snow." "White Sky."

Names as beautiful as this night.

CHAPTER 15

Between Solitude and Passion

I FASHIONED A MOUSETRAP using the salt container and some jerky. The mice had never seen a trap before, so it was easy to catch them. I caught three in the morning and half a dozen more later in the day, and released them all in the snowy field. But the mice, armed with incredible willpower, came back. I made marks on their tails before I released them, so I knew they returned within twenty-four hours.

I plugged the mouse hole. They dug another. I killed them. New ones took their place within a day or two. The elimination of an existing mouse population resulted in mouse migration from other territories, maintaining a carrying capacity of a dozen mice in the bunker. The precision of nature amazed me.

(I later made a mousetrap for the next person to stake out in the bunker. He was puzzled and then amused when I told him the story. But once he experienced what I had, he became obsessed with catching mice. Instead of trapping them, though, he requested rat poison when he ordered supplies at the base camp. The mice had no experience with rat poison or any sort of built-up immunity to it. Like smallpox wiping out the native Ussuri, poison would eliminate mice from this area altogether. I tried to talk him out of using poison, but the mice were standing in line waiting to take over the utopia that was our bunker. What were we to do? We put our heads together and came up with a plan: Each time a tiger came by, we would put a rice ball at every mouse hole. This would keep them busy and quiet for the day.)

Apart from the mice, the bunker was cozy and environmentally friendly. The air was fresh, the sounds of nature surrounded me, and I had plenty of natural light. It was a little cold and crowded—when I lay down, my head and feet touched the walls; when I stood up, I had to stay hunched over—but it was okay once I became used to it.

This was a typical day in the bunker: I would wake up early in the morning when it was still dark out. I wouldn't want to leave the warmth of the sleeping bag I had been heating with my body all night, but I would make myself get up. The cold would seep in through every pore. In the darkness, I'd find my bearings and reach for the power button on the little monitor. The screen would emit a dim light that filled the inside of the bunker. First, I would check the status of the cameras to make sure they hadn't frozen overnight or become coated in frost or dew. The cameras were supposed to withstand cold up to -45°C, but it sometimes didn't work. I would check the front and rear cameras for any nearby activity. If there were tiger or deer tracks, I would examine them closely.

Once these essential items on the checklist were ticked off, I would open the bunker entrance and allow light into the interior. The relatively warm inside air would rush out, replaced by the -30°C air outside. I would get goosebumps as though my skin had just touched a frozen piece of metal. The cold always came at me like a wall of despair and made me feel small. Ventilation complete.

The routine that followed was no different from what I did at home. If I were home, I would read the paper on the toilet and wash my face. I did more or less the same thing in the bunker. First, I would urinate. When I began this stakeout, I brought two fifty-liter tanks with me. One was for water and the other was for urine. The amazing thing was that the water tank emptied at the same rate as the urine tank filled up. Sometimes, I would have a bowel movement. I would deposit it on waterproof parchment paper, fold it up, and seal it in a plastic zipper bag. I would put this in a bigger zipper bag, blocking the smell under two layers of plastic, drop it in a large plastic pail, and close the lid. After that, I would wash my face like a cat: I'd put a bit of water on a towel and rub my face with it, then wipe my teeth with a piece of tissue paper. Brushing teeth was not allowed in the bunker because it required too much water and the smell of toothpaste could be a problem.

Then it would finally be time for breakfast. I didn't move very much inside the bunker, so two meals a day were enough. First, I'd pour some water into the kettle. If the bunker was sealed, the inside temperature stayed at -5°C to -7°C. When it was -30°C outside, however, the inside temperature couldn't help but drop as well. This meant the water in the tank would freeze solid if left on the ground. So the water tank was placed in a water storage unit, a hole in the ground lined with Styrofoam, and I would transfer small amounts of water to a bottle for daily use.

Next, I would pull out the can of butane gas I kept warm in my sleeping bag like a hen brooding an egg. Gasoline burners work well in cold weather conditions, but they smell, so we used butane burners instead. Butane is scentless but freezes easily below zero, so I always slept with the can I planned to use the next day. I would shake the can, attach it to the burner, and click the igniter, and blue flames would spring up. I would put the little kettle on the burner to boil water, then pull out a frozen rice ball from a plastic bag and drop it in the boiling water. I assemble two hundred to three hundred rice balls individually wrapped in plastic bags before I begin a stakeout. They freeze solid, so I don't need to worry about them going bad.

When the rice ball defrosted and became warm, I would scoop it out of the kettle and eat it, washing it down with the remaining water. I drank green tea, rice tea, and other teas that don't smell. I had to always be careful not to let the smell escape. It was the greatest hazard to bunker life. I would have jerky, seaweed, and salt with my rice and eat dried fruit and nuts for dessert. The food was plain, but I was already sated by being in nature. I would eat as the sun rose over the East Sea. Mealtimes were always fun.

At 10 a.m. sharp, I would contact the base camp via radio. To prolong battery life, we communicated only on predetermined days every few days or few weeks, unless we had an urgent message. Because reporting on the movement of tigers was our first priority, our communications would always begin with news of the tigers. Next, I would update them on my food and battery supplies and order any items or equipment I wanted to include in the bimonthly supply run. Only then would we catch up on each other's lives and exchange other news.

Between 11 a.m. and 1 p.m., I would take a nap. Tigers don't like to be on the move in the middle of the day and usually nap during these hours as well. I would sleep even if I wasn't sleepy so I could be half-awake at night and hear what was going on around me.

I would take some time for my hobbies from 1 to 4 p.m. I would read, meditate, write in my journal, or play with the mice to relieve the stress of bunker life. Once I started to feel trapped, I knew a mental breakdown could be close behind. To keep my emotions steady, I had to relax and enjoy myself as much as possible.

I'd look through the viewfinder from time to time as I worked on my hobbies. Through the viewfinder of the camera, which was always pointed at the same spot from the same angle, nature appeared as motionless as a landscape painting. But if I looked closely, I could see the living landscape slowly moving and changing. The longer I observed, the livelier the turn of the seasons. A close view of one spot in nature is a joy only bunker life can provide.

When a wild animal appeared on the screen, I would feel as happy and excited as I did when a tiger came. One day after a snowstorm, an eagle searching for a place to rest landed on a nut pine. He dusted the snow off the branch with his wings and tapped on it a few times with his talons, then shrugged his wings. With an expression on his face that seemed to say *Life is too much sometimes*, he became a part of the snowy scene. I saw the burden of life in the eagle.

At 4 p.m., I would have an early dinner and observe the area more closely. This was the time of day when tigers became active. Due to battery lifespans, I couldn't leave the camera on all the time. But between 4 and 5 p.m., I would turn the camera on every ten minutes, and then every five minutes between 5 and 6 p.m. From 6 p.m. until sundown, the camera would stay on.

When the sun went down, I would switch to the night camera lens. I needed good light to get a decent image with this lens, and the moonlight reflecting off the snow provided just the right amount. If there was a full moon, the reflection became even brighter. When a tiger appeared in this snowscape, I felt a fantastic chill running down my back; the forest looked frozen by the brilliant light.

When the light wasn't good and I decided to forgo the night shoot, I'd pull down all three layers of the camouflage blankets and board up the entrance with the wood panel. The bunker would become as dark as the abyss. I'd lie still and listen to the sounds coming from outside. The never-ending rhythm of ocean waves and the wind sweeping its hands over the mountain ridge, the sound of gulls heading home late at night, the owl couple calling to each other, and a fox barking in the distance—these were all sounds that reminded me of the solitude of night.

My thoughts would stretch on and on. I would think of good people and bad. I'd get angry thinking of the people who never contribute but jump at opportunities to take credit, but my anger seemed silly against the sound of the crashing waves. I would mull the things I wanted to do when I got out of the bunker. *I should be good to my parents while they're still around. I should look up a friend I haven't seen in years.* As my thoughts skipped from one thing to another, my imagination would take flight. I would forget where I was and revisit memories from my childhood.

Then I would doze off. Charged with anxiety that a tiger might appear, I'd dream in a state of half-sleep. I often dreamed of tigers: *The snow comes down hard. White Moon and White Snow play in the snowy field. White Sky nips at his mother's tail. The tigers nuzzle and lick snow off each other. White Snow takes off and everyone runs after her. They disappear into the snowy Ussuri forest.*

I'D BEEN ON stakeout for over two months. I was almost out of batteries and rice balls. I needed a fresh supply of drinking water and had to dispose of my waste. Fresh supplies were supposed to have arrived two days earlier, but they'd been held up because of the weather. The base camp could deliver supplies to bunkers on the coast by boat, but there were no docks on the beaches here and it wouldn't take more than a medium-size wave to make unloading dangerous. So if there was even the slightest hint of waves, whoever was on delivery detail came over the mountains on foot via Mayak Village, ten kilometers south of the bunker. Climbing several mountains with the heavy batteries and supplies took great determination.

Bloody Mary traveled here from Mayak Village several times a year. When I was on delivery detail in previous years, I would suddenly feel tense or cold as I walked down the forest path that Bloody Mary herself had trod upon. I would stop at the slightest rustle to make sure the coast was clear, and when a gust of winter wind swept up the mountain from behind, I couldn't help but look back to make sure there were no tigers behind me.

When supplies arrive at a bunker, the most important thing is to avoid being seen by tigers. The ideal delivery time is a week after a tiger visit. If the delivery person arrives too soon, tigers might still be in the area, and if they arrive too late, a tiger might be returning. The ideal window of time is the one to two weeks when the tiger has surely cleared out but hasn't had time to make its way back.

The supplies finally made it. After being in the bunker for so long, I felt emotional at the sight of a colleague arriving. We looked at one another, each moved by the sight of the other, but didn't have much to say, or much time to say it. In case there was a tiger on its way, I quickly lowered the supplies into the bunker and gave my colleague the used batteries and waste, and he left. Had we been absolutely certain there were no tigers around, we would have sat on a sunny hill and chatted for a bit. We'd have exchanged stories we couldn't over the two-way radio.

I was alone again. I felt blue. I was reminded of the fact that humans, unlike tigers, are a species that live in groups. My arms and legs felt weak, and my chest felt tight. The bunker suddenly resembled an endless desert or a bottomless well, and loneliness washed over me like a tide. The two-square-meter space that surrounded me made me tremble with claustrophobia. I lay shaking in a state of despair until night fell.

Eyes closed, I was as lucid as if my eyes were open. I still had hours to go before sunrise. If I could, I would jump into the ocean and cry my heart out. It was not the fear of beasts or the cold, but loneliness that paralyzed me in the bunker. How were tigers so fond of being solitary? The anxiety of being alone and the desire to wait for the tigers pulled me in two directions. I thought about who and what I was enduring this loneliness for and how I would face the world if I were to bolt out of this bunker right then. I consoled myself and pulled myself together.

The next morning, it was snowing. I tried to cheer up by telling myself that without a small but cozy space like this, I would have no chance in a snowstorm. I repeated this to myself like a mantra: after this snowstorm and a few others, I'd be back at the base camp warming myself by a wood stove, drinking vodka, and falling into a deep, restful sleep. The snowstorms saved me from my solitude.

Time passed, and I was back to appreciating the comforts of the bunker. Humans are endlessly vulnerable in nature, but when I was gazing out of my small space, a snowstorm in the uninhabited wild made pretty good scenery. The stronger the wind and the wilder the snowstorm, the more the bunker felt like a warm, cozy bedroom. People wonder why I would want to be in a filthy, uncomfortable hole like that, but its warmth was like a drug to me.

A rolling stone gathers no moss. But a rolling stone also wears down. In our effort to live life to the fullest, we sometimes deprive ourselves of the time to reflect on what is truly missing in our lives and how to fill that void. In life, it's best to be a mossy stone sometimes. Without taking the time to stay put and sink deep into the ground, you can lose your way.

CHAPTER 16

Merry Christmas, Bloody Mary

IT WAS STILL snowing two days later—a snowy kingdom as far as the eye could see. I kept watching the snow fall on the mountains as I waited for the tigers. I thought I saw the orange fur of a tiger knocking snow off the bushes as it emerged, but it was just more snow blurring the line between ground and sky as it fell.

"This is Diplyak. This is Diplyak. Nothing to report. Merry Christmas!"

"This is Petrova. Tiger tracks found between the Deer Valley and the Coastal Ridge. Be advised. A merry Christmas to you, too!"

I was communicating with Petrova Lodge, the location of one of our base camps. The tracks of three tigers had been found inland two days earlier, thirty kilometers northwest of the Petrova coast. They must be White Moon, White Snow, and White Sky. It was a day's trip from there to here for them, but I couldn't predict when they would arrive. What were the tiger siblings up to in the snowy forest? What were my kids up to? When I was home, I longed for the mountains. When I was in the mountains, I was homesick. I had been repeating this pattern for fifteen years. I was weary.

Four days later, I heard a deer cry in the middle of the night. The deer, which had been making warning calls to other deer, suddenly squealed. The monosyllabic cries stopped after two or three calls. Something was going on. Something seemed to have attacked the deer. A lynx living on the coast? Or were the tigers back?

I scanned the field from left to right through the night-vision lens. I saw trees and shrubs buried in snow. I carefully studied the far side of the snow, the green pine needles, the dry branches, and the faded shrubs. There's a limit to how well you can capture the tension and shock of seeing a tiger in the Ussuri wild. The footage gives you only an inkling. I wondered if I'd be able to capture so much as 10 percent of it.

The Diplyak coastal basin was still. I felt some sort of energy spreading in the stillness, but I couldn't identify the source of it. In moments like this, it's wise not to move the lens. The source might be looking right at you. I stopped moving and waited with the camera on standby mode.

The next morning, I woke to the sound of a herd of deer galloping by. Ussuri deer had been traveling down into the basin since the previous day. I had seen a few go by since the stakeout began, but this was the first I had seen herds migrating like this. The cold was really starting to set in, and the recent snowstorm had likely forced the animals to migrate. Deer tend to move in search of food after big snowstorms. Diplyak is warm because it faces south; also, it's easy for deer to find acorn and leaves in the oak forest and forage for marine plants on the beach.

From behind, the sound of hooves drew nearer, and a deer leapt off the roof of the bunker and landed right in front of the entrance. With its white, heart-shaped bottom right in my face, it let out a warning call. It was a strange feeling to see a deer warn the others about something when the very thing it should beware of was right behind it. The deer pricked its ear and cocked its head sideways before kicking the ground a few times and rejoining the herd running through the basin.

We had picked the wrong spot for the bunker. It was only after we had built it that we'd noticed we were right in the middle of an animal path. We should have paid more attention when we were drawing the eco-maps.

Cold winds brought a mountain of gray clouds, and on the horizon, the snowy coastal ridge bled into the gray. When it's overcast and windy, tigers move earlier in the day. That day, I was determined to stay focused during the day as well. If the squealing deer the night before was the doing of a tiger, it would be back today. When tigers catch deer, they drag their kill off to a secluded place and feed on it for two or three days. They find a warm place to rest after they've had their fill and return when they're hungry again.

I knew a few places nearby where the tiger could be. From the bunker, the beach was to the left and the young pine grove was about 150 meters to the right. In the summer, I'd once seen traces that Bloody Mary's family had left behind after dragging a deer there to eat. There was a small brook behind the young pine grove. The tiger must have eaten its kill in the pine grove or by the frozen brook, or maybe in the knoll behind the brook. Long ago, I'd seen a deer carcass on that knoll as well. Besides, the short hazel trees and young oaks on the knoll that hadn't shed their leaves completely would make for a good place to hide. If it was really a tiger who had done the hunting the previous night, it would have dragged the deer to one of those three places. Perhaps it was already near the forest to scope out the situation. Once it occurred to me that a tiger might be nearby, I was reluctant to adjust the camera lens.

At three in the afternoon, I aimed the lens at the right side of the forest and slowly panned back to the foot of the hills. I saw bushes sitting under snow and the green pine grove. Big, sturdy oaks stood along the ridge, and young oaks barely one meter tall crowded the foot of the hill. I was scanning the grove of young oaks when I suddenly stopped. It had been there all along! I didn't know how it had crept up so fast, but there it was, its whiskered face peering out from between the yellow oak leaves fluttering in the wind. Its thick torso was obscured by the young oaks, and its striped face was expertly camouflaged by the yellow leaves and brown branches.

I let out a sigh of relief. I'd have missed it altogether if it hadn't been for the stripes. Or rather, I'd have given myself away by moving the camera lens. Fortunately, it wasn't the lens the tiger was looking at. It stayed camouflaged for over ten minutes as it surveyed the area. To the left, waves foamed as they crashed on the shore, and to the right, a bare ridge embraced the forest in the small basin. Behind the tiger, valleys and ridges cast a sharp contrast between light and dark as their bold lines climbed toward the peak.

Out of this scenery came the tiger, slowly making its way into the clearing.

The tiger sat primly like a housecat and started to gnaw on the leftover deer from the day before. Its black-striped orange fur was beautiful against the blue waves of the East Sea. If tigers in the mountains look proud and powerful, a tiger by the sea looks carefree and gentle. The size and bone structure of the tiger indicated it was a female. Her expression was too unguarded to be Bloody Mary, so she must have been either White Moon or White Snow. I was leaning toward White Moon because she was traveling alone. During the summer expedition research, we'd often found this smallest of Bloody Mary's cubs wandering on her own. Like children born into poor families who have to grow up fast, White Moon seemed to be preparing to part with her family sooner, since there hadn't always been enough food for the youngest of the litter.

In daylight, she looked bigger than she had when I'd seen her at night some time earlier. If it's difficult to tell a mother and

ABOVE Bloody Mary's footprints on the beach.
When I found footprints like these, I wished I could follow them forever.

TOP LEFT Bloody Mary's vast territory ran along the Sikhote-Alin Range and stretched inland. Her family played somewhere on the Dragon Spine far in the distance.

BOTTOM LEFT Diamond dust. The microscopic specks of ice glitter in the sunlight as they float through the air and land on the field of snow.

ABOVE In the early spring of each year, Ussuri deer that have survived the harsh winter pass through the Basin of Skeletons on their way to the coast.

TOP The Ussuri believe that the soul of a person comes from trees, and that when they die, the soul of a man returns to a willow, and the soul of a woman to a birch.

ABOVE Each year at the end of October, it's as if the door to a massive ice cave opens and cold winds rush out from within. The rivers freeze solid, the forest is buried in snow, and the six months of winter begin.

ABOVE One arm of the Sikhote-Alin Range runs along the coast to form a coastal range. It was a good place for Bloody Mary to raise her cubs.

TOP The Great King drew back his lips to reveal his fangs to me. It was a silent warning not to do anything foolish. The gesture drained the last bit of courage I had in me.

ABOVE Experienced tigers don't mark their territories on the ocean cliff during high tide when the urine can easily be washed away, but young tigers spray here.

TOP LEFT Tigers display affection between mothers and cubs, fathers and cubs, and mated couples.

BOTTOM LEFT The cubs hide in the forest while their mother hunts. Wild tigers leave the den with their mothers when they're two months old, and begin a lifetime of wandering like the wind and water.

ABOVE The Udege wander all over the Sikhote-Alin Range gathering ginseng roots. The last animists in Ussuri, they worship the tiger and call it Amba.

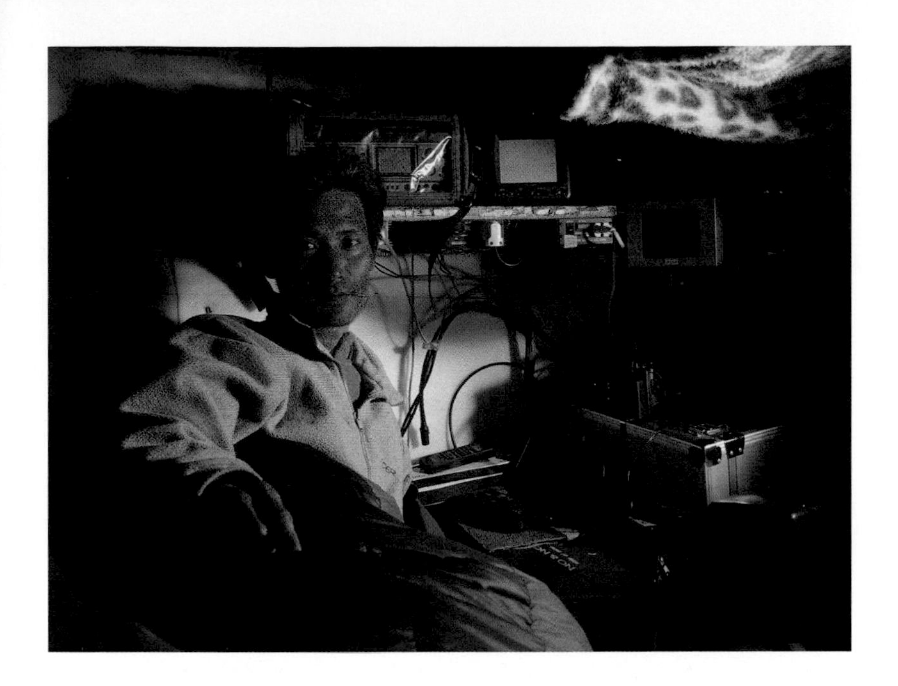

TOP LEFT A fully grown cub marked his territory on the coastal cliff with urine.

BOTTOM LEFT The mother tiger traced her cub by his smell.

ABOVE I waited in the cramped underground bunker for the tiger that never comes.

RIGHT After staggering through the snowy fields, Bloody Mary fell. She lay and dug at the snow, and took her last breath with the strength that remained after she could no longer kick her legs.

ABOVE The bunker after Bloody Mary's attack. She was probably up on the range somewhere, looking down at me. I wouldn't expect anything different from an Ussuri tiger, much less Bloody Mary.

TOP LEFT I saw calm control in every move Bloody Mary made to protect herself and her family in the brutal fight for survival.

BOTTOM LEFT The wild tiger lifted her tail and sprayed the large tree. Urine is a mode of communication for tigers and a powerful perfume female tigers use to attract males.

ABOVE Siberian tigers don't generally venture into wide-open places in broad daylight, but they're sometimes spotted on Apasna Beach, which is closed off by steep, dangerous rock cliffs.

ABOVE Flehmen response. When tigers return to a place they've missed, or when they feel they are in a spot as safe as home, they make this face as they sniff the air.

cubs apart by their size, it's a good idea to look at their faces. You can tell from their expressions if they're old and aloof or young, innocent, and inexperienced. You can also tell the difference by how cautious or clever they are. Bloody Mary would never move around in broad daylight.

A deer made a warning call in the forest to the right. White Moon's gently waving tail suddenly stopped. Her strong front paws planted firmly in the ground, she looked in the direction the sound had come from. Nervous due to the time of day, she got up on her hind legs. Then she looked away from the forest and walked off into the oak grove she had come from earlier. The yellow leaves rustled as she walked by, as though brushed by a gentle breeze, and then grew still again. She'd been fast. It was a brief encounter, but I'd managed to get this beautiful moment in the wild on film. On the other hand, I was worried that she'd left because she'd picked up on my scent or the movement of the camera.

The snowflakes began to fly again. I wouldn't be able to see the bright, full moon floating in the sky that night. When I thought about it more, I realized Bloody Mary and her family tended to come to the beach around the time the full moon was out. They had been there on November 27, four days before the full moon, and now on December 29, one day before the full moon. When we'd researched this beach in August, we'd found their tracks right after the full moon at high tide, and this pattern was true in the past, too.

What was the reason for the tigers heading to the beach for the full moon? It may have had something to do with the steep cliff along the beach and the waves. Climbing the dizzying cliffs of the east coast on a moonless night is not a welcoming thought, perhaps even for tigers, but not as unappealing as moving around in daylight. Wide-open beaches are sometimes unavoidable when traveling along the coastal ridge, and being in an open area during the day is one of the wild tigers' least favorite things to do. So they wait for bright, moonlit nights to travel along the coast.

Also, two to three days before the full moon is the ideal time for tigers to hunt in this area. During a full moon on high tide, when the moon's gravitational pull creates the greatest tidal range, waves as tall as houses whip the beaches and fill the coast with a deafening roar that helps tigers approach their prey undetected. The wind eddies following the big waves also make it difficult for the ungulates on the beach to smell the tigers. The more I watched Ussuri tigers, the more I was amazed by their intelligence. Every move they made was careful and calculated.

The snow stopped and the clouds parted as evening fell. It had been a few days since the winter solstice, and the days were very short. Between 6:30 and 6:40 in the evening, the camera hit its exposure limit. I was debating whether I should swap in the night lens, hoping to hold out just a little longer, when I saw the hazy silhouette of a large animal in the distance. It was 6:43 p.m., right after the day lens stopped being usable. I felt around the dark bunker and quietly put in the night lens. Thanks to all the practice, I was used to handling equipment in the dark. I turned the camera on and unlocked the tripod joints. The viewfinder emitted a foggy light as it came on. I panned the camera to get the angle as quickly as possible without catching the tiger's attention. A pair of eyes flashed in the dark, and I saw stripes. A brawny tiger was casting his gaze around. I looked straight at him without hitting the record button. The moment he turned to look at the mountains, I hit record. I heard the quiet *whirr* of the tape locking in place and starting to turn. The sound seemed exceptionally loud. I slowly zoomed in and found the focus. The image of the tiger became clearer. It was White Sky. He was still gazing back at the mountains.

White Sky headed in the direction he was looking. I followed him by moving the frame diagonally. And then another tiger entered the frame. It nuzzled against White Sky's neck and snuffled affectionately. White Sky caressed the other tiger's neck with his tail and walked off again. Yet another tiger came into the frame before long. All three rubbed noses and nuzzled against each other in greeting, congregating under the full moon. It was White Moon, White Snow, and White Sky, five days after their tracks had been found thirty kilometers inland. They must have followed the deer here. It was time for them to become independent, but these siblings were still traveling together. They played together, nudging each other, snapping at each other's tails, and frolicking.

From behind the bunker, I heard footsteps approaching, crunching on snow. The hairs on the back of my neck stood up. The heavy sound of a large cat's soft paws on snow... Bloody Mary?

The crunching got closer and closer until it moved on top of the bunker roof. The roof creaked. My heart began to race and I couldn't breathe.

Calm down. Calm down.

It all happened so fast that my heart was ready to jump out of my chest, and I was dizzy in spite of my effort to calm myself. Bloody Mary was right above my head. *One, two, three, four...* Thirty seconds went by. It felt like an eternity. Bloody Mary suddenly leapt over the entrance of the bunker, and the camera caught her as she walked into the thicket, brushing snow off the lower branches in her wake. With her tail held straight up to avoid scratching it on the branches, her body slinked from side to side as she walked confidently up to her cubs. Bloody Mary had followed the deer that had passed through in the morning. I chastised myself once again for building a bunker right in the middle of a deer path.

I caught all four tigers in the frame. As I guessed, it was Bloody Mary, the matron of the Basin of Skeletons. She had finally appeared before me. Her long tail dangled behind her with just the tip turned up like a snake's head. Her torso was slender and firm, her shoulder blades jutted out in a sharp angle, and her strong head was held up high. She was tidy, disciplined, and clean. I felt power and unyielding will in her. She was calm and reserved, like a fighter who calculated her every move to protect herself and her family in the brutal struggle for survival. She radiated a wildness that could only come from fighting through life alone, without relying on anyone else.

The cubs snuffled affectionately at their mother as they nuzzled against her neck. But Bloody Mary did not reciprocate. She must have sensed that something was amiss, for she was busy sniffing and surveying the area. I carefully zoomed in until Bloody Mary's face filled the frame. The look on her face was not one of a loving mother. She was on edge, full of suspicion and caution.

She was looking around when, suddenly, she spotted the lens. The lens froze and Bloody Mary froze as well, but she did not take her eyes off of it. Her gaze stayed riveted, as if glued. My heart rate suddenly spiked. I had been too careless in my excitement at seeing Bloody Mary for the first time. I had forgotten that she was an exceptional mother. I should have been more careful with a veteran like her.

With her glare fixed on the camera lens, Bloody Mary slowly turned her body around. She had the look of a mother tiger who would take on the entire world for the sake of her cubs. Her long, serpentine tail held straight up, she walked through the frosty thicket, straight toward me. A pile of snow on the elm branch above the bunker plopped down as the wind blew. She came closer, her eyes staring straight at the lens. The close-up image of her in the viewfinder turned blurry and then disappeared altogether. I couldn't see the look in her eyes anymore. I couldn't zoom out either. I held my breath and stayed very still.

The *crunch crunch* of her paws on snow got closer. My blood was freezing in my veins. I wondered if I should retract my lens right then, and realized that it was already too late for that. If I got caught like this, everything would go to waste. The bunker would be destroyed and my life would be in danger. With my left hand on the focus ring of the lens jutting out through three layers of blankets, and my right hand on the tripod handle inside the bunker, I held my breath. I left everything to fate and hoped she would pass me by.

Crunch. Crunch.

Bloody Mary continued toward the entrance. One, two, three, four steps. Stop. She was on the left side of the lens.

Bloody Mary sniffed the lens from the left and right. I had taken off my left glove to manipulate the sensitive focus ring, and now I felt Bloody Mary's warm breath on the back of my left hand. I was so tense my back was about to break, and my skin was covered in goosebumps. Her breath and her stiff whiskers grazed the back of my left hand. The back of my hand spasmed. And then her strong front paw struck the camera lens. The lens bent and the microphone snapped.

She let out an indescribable, blood-curdling growl, and the attack began. She caught the zoom cable that was dragged out with the lens when it fell off and yanked at it. The copper wire inside the cable cut the back of my right hand as the cable snapped. A sharp pain shot through my body. Bloody Mary clawed off the camouflage shrubs at the entrance and started to rip off the wooden paneling. The cubs rushed to her side. They used their spade-like front paws to dig up the soil around my hideout. Small holes appeared in the corner of the bunker. They stuck their noses in the holes and sniffed. I heard the raspy breath of a tiger right by my ear.

One tiger got on the roof of the bunker. Then another. Soon, I heard the sound of the pine boards snapping. The tigers were digging up the earth on the top of the bunker. I broke out into a cold sweat and trembled all over. Of the three layers of blankets at the entrance, one had already been ripped off and the second was about to come down as well. The roof bent and started to cave. The four tigers did their worst while the helpless man sat paralyzed in a ditch.

A third tiger climbed on top of the bunker, and the roof broke with a loud crash. Frozen dirt and clumps of snow fell into the bunker along with a tiger's back foot. My mind went blank. The tiger must have been equally stunned that its foot had fallen into a dark hole, for it kicked frantically to pull itself out. The kicking broke the pine board a second time, and I heard the tigers running away.

Once the paw retreated from the bunker, I suddenly snapped out of it. I thought about getting up to push the pine board back into place, but I stopped. As hard as it was, I had to keep my wits about me to the end, and I sensed that this was not over yet.

The tigers were quiet. So determined just moments ago to scratch and tear and sniff and find out what was in the ground, they had instantly grown still. The footsteps grew distant and then ceased altogether. Silence had returned. Not a mouse in sight. Only the occasional plop of snow falling from branches.

CHAPTER 17

By the Skin of Tiger Teeth

BLOODY MARY HAD not left. The footsteps moved away and suddenly stopped. She lay quietly hidden somewhere, spying on me. She would be sitting in the sphinx position, with her eyes on me. But I couldn't tell where she was. The cubs were as still as their mother.

I instinctively knew this was a real crisis. Whoever moved first would lose. Just in case the situation worsened, I carefully boarded up the entrance with the panel and pulled out the cannon flare to wait. Unless they came into the bunker, I wouldn't use the cannon flare. I'd made it this far, so I figured I might as well keep up the ruse.

Twenty minutes went by without either man or tigers making any move. Frighteningly cold wind poured in through the collapsed roof. I wanted to rub my hands together, but I couldn't manage to lift a finger. I was getting colder faster because I couldn't move. The tigers frightened me, but the brutal power of physics, the temperature of -30°C, was becoming more worrisome. I was freezing from head to toe and sat as stiff as a rock.

Thirty minutes later, I heard footsteps slowly approaching.

Crunch, crunch. The squeak of paws on wet snow was as unnerving as an awl piercing my brain. Bloody Mary had indeed been cunning and tenacious enough to lie in wait just behind the bunker for thirty minutes, testing me.

She came by the entrance again for a sniff. I thought I felt her breath on my left cheek. I was relieved that I'd had the presence of mind to block the entrance during the brief time she'd been gone, but I was worried about the collapsed roof. What if she came in that way? Knowing tigers, I didn't think she'd venture into a dark hole without knowing what was inside. But what if one of her silly cubs were to fall in? If I was trapped in this space with a tiger, I would have nowhere to run. The tiger, also with no place to run, would instinctively attack me. I prayed that this would not come to pass.

The cubs also circled and sniffed the bunker. I could hear sniffing coming from all sides. They were all grown up and as big as their mother, but they still did only what she did. They stopped when she did and started moving again when she moved. Fortunately, none of them climbed on top of the roof. They must have been scared too, remembering how the ground had given out and swallowed a paw.

After all that poking around, there's no reaction from inside. I saw something move and I smelled human, but there's nothing... Bloody Mary must have been confused now. What if I had moved, thinking she was gone? I calmed my racing heart in relief.

The attack resumed. They scratched, bit, dug, and made a huge racket before the sounds suddenly stopped. All I heard were cold blasts of wind coming in through the collapsed roof.

Almost an hour later, Bloody Mary came over again. She sniffed and scratched. She pawed and dug like she was intent on getting to the bottom of this mystery. This went on all night.

At 4:10 a.m., I heard cubs wrestling with each other and running about. They seemed to have let down their guard. Then I heard them leaving, one by one. Instead of just pausing, the footsteps receded.

If Bloody Mary had figured out what the bunker was, she would never return again. But it seemed she was still suspicious. Based on the fact that she didn't leave until dawn, I believed she still thought there was something fishy about the bunker. There was a good chance she would spy on me from the top of the mountain and return at night for another examination. This was how any Ussuri tiger would behave—not to mention this was Bloody Mary we were talking about.

It was -34°C. I was so cold. Just as it is darkest before the dawn, it is also coldest before the dawn. The towel I'd hung up the previous morning after wiping my face had become stiff as a board, and my apples were frozen so solid that I could use them to hammer nails. There was frost clinging to my nose and moustache. After nine hours of sitting very still, I couldn't move. I'd experienced colder weather before, but the chill I now felt was much stronger. Now I understood how quickly your perception of temperature drops when you can't move. I wanted to crawl into my sleeping bag, but I waited a little longer just in case.

I knew that, years later, when I looked back on that day, I would hardly believe it had happened. The very thought of it turned my blood cold. The fragile pendulum of life had swung; death had come right up to my nose and retreated. Once the swinging between life and death stopped, when the uncertain vibration of the moment was gone, it was difficult to express exactly how it had felt.

Dawn began to break. It was the last day of the year. I wiped the congealed blood off my right hand and cleaned up the mess in the bunker. I pushed the roof up, and clumps of dirt and ice fell inside. I propped up the roof with the tripod. I swept up the dirt and snow and cleared the bed. I rehung the blanket over the entrance and checked on the wood panel door as well. Fortunately, the antenna cable had been left intact.

At 10 a.m., I reported the incident to the base camp and warned them not to approach. I thought about Bloody Mary, who was resting on the mountain with a good view of the bunker. I prepared for the night by making tea and having a simple meal of a melted rice ball. I continued to face the challenges of my life here, but I felt a little sad. I hung my head in spite of my efforts to put on a brave face. It took a while before I could pull myself back together. When I felt calmer, I sealed the bunker's entrance and crawled into the sleeping bag.

Night came again. The winds were calm and it was still all around. The sky had been clear all day, so the full moon must have been shining down. A Eurasian eagle owl hooted in the distance. I pictured the eagle owl sitting on top of a coniferous tree, its wide, red eyes searching the snowy forest for prey. The mice had returned to the bunker, but were still cautious after the previous night's sneak attack.

It was after midnight and Bloody Mary hadn't returned. I was disheartened, as I thought she had found me out. My heart felt empty, and I was feverish. I dropped off to sleep.

Three a.m. The sound of footsteps approaching woke me up. I opened my eyes. The audio level on the sound monitor was rising and falling.

Crunch. Crunch. The careful footsteps on snow circled the bunker once and then came toward the entrance. A low, metallic growl came vibrating from the tiger's throat. I felt Bloody Mary's breath through the one remaining layer of blanket and the wood panel. Now that she had come, the emptiness and disappointment of never seeing her again instantly vanished and I was horrified again.

Unlike the day before, I was better able to gauge the situation. They say you can survive a tiger attack if you keep your wits about you, and this is true. Now that I was calmer, I could tell where Bloody Mary was standing even when her footsteps stopped. The cubs were coming down the hill behind the bunker. I was fully prepared. I decided to use the tripod to hold the roof firmly in place instead of filming. The entrance had been sealed as well. I had crawled into the sleeping bag early, so I wasn't cold, either. A strange feeling welled up inside me, like the sense of relief you feel when you're sitting cozily by a fire listening to a snowstorm rage outside. The sentimentality of the morning had been replaced by the thrill and tension of the swinging pendulum again.

When Bloody Mary didn't hear a single peep coming from the bunker, she walked off to the right. The cubs followed. I carefully turned on the security camera to find her looking back and waiting for a cub that was lagging behind. It almost looked like she was glaring at the bunker. Bloody Mary stalked off, and White Moon, who had been grooming herself in the bushes, quickly followed. Bloody Mary's family disappeared into the pine grove. It was ten minutes to dawn.

It appeared Bloody Mary was less suspicious now, even though she had lingering doubts. I had fooled her. Compared to the day before, when I had spent hours in fear of the roof collapsing and being attacked by four beasts that weighed over two hundred kilograms each, this day was... I wished I could say it was better, but it was just as rough.

The sun rose over the East Sea. It was powerful and red, as if to signal the beginning of a new year. The skies were clear. Had Bloody Mary left?

I decided I would film again that day. I slept during the day and started preparations at 4 p.m. Because the temperature inside the bunker had dropped dramatically the previous day, though, I ran into problems with the batteries; they were discharging quickly. I turned on the security camera using what energy remained in its batteries. A tiger must have destroyed it, for I wasn't getting any image or sound feedback.

It grew overcast in the afternoon. I checked the viewfinder every ten minutes, and then after 5 p.m., every five minutes. Five-forty-eight p.m. A striped pattern flashed in the pine grove. It was the three cubs. No Bloody Mary. The cubs lingered in the pine grove, reluctant to be out in the open when it was still bright out.

After loitering for a while, the youngest, White Moon, turned and headed inland along the mountain ridge. White Sky and White Snow bravely ventured out to the shrubbed area and climbed the perilous coastal ridge. White Sky was fearless. Halfway up the steep rocky mountain, he turned only his neck and stared at the bunker. I saw the two tigers' black stripes and ginger fur appear intermittently through the cliffside pines until they disappeared over the snowy mountain ridge.

The next day, the tigers did not come.

The day after that, I contacted the base camp, left the bunker in the coastal basin, and moved into the bunker on the beach. The wind was strong, and snowflakes started to flutter. The white-haired witch had flown over from Siberia. I wished the witch would wipe the memories of my difficult encounter with Bloody Mary and my own vulnerability clean. A violent snowstorm was coming. It began to snow heavily.

CHAPTER 18

Tiger for Sale

PEOPLE IN THIS part of the world are poor. Most farmers grow little more than potatoes, and a few chickens are all they have for livestock. Very few can afford to raise pigs and cattle, so they harvest wild vegetables and fish instead. They came to the base camp bearing wild vegetables, fish, or eggs in their baskets and quietly peeked in. They wanted to sell them to us. In Kievka and the surrounding villages, it was easy enough for the villagers to exchange goods, but difficult to get money for their wares, so we tended to buy things from them for more than they were worth.

The villagers envied our equipment, cables, and other supplies. They looked at them admiringly, mumbling to themselves about what they could do if they only had this or that. Some villages didn't have electricity despite having had posts installed because they couldn't afford electric lines. Since the switch from socialism to capitalism, the wealth had not trickled down to these remote areas.

One day, a resident from one of the nearby villages came to the base camp to visit and showed us a plug. "Look at this plug. It's perfectly good," he said. "If we can wire it, we could use electricity."

We couldn't help but give him some electric line. We handed him a spool of it as a gift. News spread to others that the village of the man who showed us the plug had electricity thanks to our help. A few days later, another man came and handed us a piece of paper. Here's what it said:

10 pink salmon
10 bundles of Alpine leek
3 bottles of five-flavor berry wine
1 sack of potatoes
5 kilograms of chaga mushrooms
10 kilograms of oak charcoal
2 cubic meters of firewood
P.S. I will tell you if I see any tiger tracks.

He said that he would give us everything on this list. In exchange, he wanted a saw. He was a charcoal maker who made charcoal out of oak, and he needed a saw for his livelihood.

"How does a charcoal maker not have a saw already?" I asked.

He mimed sawing and said, "*Ratatata, ratatata, ratatata*... All done."

He did own a saw, but was dying to get a chainsaw. I understood the envy he must have felt as he toiled, while others with chainsaws brought down big oaks with just three swipes of that tool. He knew he had to give us something in return for the chainsaw, but he didn't have anything to offer us. That was probably how he had come up with the idea of bringing us an inventory of everything he could find in his house. So we had to get him a chainsaw. Such exchanges built a bond between us and the villagers. When they saw tiger tracks while they were out working in the woods, they always let us know. They became our allies.

Things could get very hectic quickly when we planned a trip to Vladivostok. To the residents of Kievka, a trip to Vladivostok, four hundred kilometers away, was a truly big deal. First, they had to walk to Lazo to catch a bus that ran only once a day, which meant they would have to spend the night in Vladivostok. Most of them couldn't afford to stay at an inn and didn't know anyone in the city to take them in for the night. Since we had a jeep, we wound up taking care of all the business the villagers had in the city. We handled everything from buying supplies to delivering mail. When word got around that we were going into the city, villagers would hand us letters and give us approximate locations of where the recipients lived. No matter how early we left in the morning, it was always past midnight when we returned after doing all our chores.

When we were about to set out, children flocked to the jeep, their eyes brimming with desire to visit the big city. When they looked up at us like that, we couldn't say no. When we drove down the dusty, newly paved road in the jeep loaded up with children who were in a motor vehicle for the first time in their lives, I felt nostalgic for my own childhood.

Years ago, I got to know a boy named Maxim, the son of a forest ranger who once worked with us. He had an innocence about him and was too shy to talk to strangers. Once, when he was in high school, we found him gazing longingly at our jeep as we were getting ready for a trip into the city.

"We'll take you if you get your parents' permission."

He was overjoyed to hear this, and shortly returned with his sister, who was one year older than he. She was a very pretty girl. We went over the hills, raced down the newly paved road, and got skewers from a vendor by the brook for lunch. After the big lunch, the siblings fell asleep in the car, leaning against each other. They were still asleep when we arrived, and their eyes grew as wide as saucers when they awoke and looked around with wonder at a city for the first time. Feeling sorry that they couldn't afford to enjoy themselves, I gave them fifty dollars for spending money. Their faces turned red with pure joy and gratitude as they thanked me.

Again, I was reminded of my own childhood. I was grateful to them for taking me back.

When I met Maxim again three years later, he was so happy to see me. He'd graduated from high school and married young. He had also been in the army, where he'd served as a motor transport soldier, so he offered to drive us around in the jeep for free. Stefanovich interviewed Maxim and hired him as our assistant. Today, Maxim drives the Ural, our traveling base camp.

Maxim's older sister is studying biology at the prestigious Saint Petersburg State University. I saw her again in St. Petersburg when I was on a trans-Siberian trip the summer we hired Maxim. We hadn't seen each other in several years, but she greeted me as if no time had passed. She had changed a great deal and turned into a very elegant Russian lady. I felt a little strange around her. Some parts of the girl I had met in the village years ago remained, but there was a small awkwardness between us. Well aware of her father's financial situation and how hard it must be to make ends meet far away from home, I gave her some money before we parted.

The Russians say all it takes to make a friend is vodka. But this isn't true. You also need time. Working with the villagers for long periods of time and bonding with them was what it took to become friends with them. But once you became friends with people from one village, it was easy to befriend people from other villages as well. News of what we were doing in the area got around from village to village. Over vodka, they introduced us to people from other villages by telling them, "It's important what they're doing here. They want to leave records of the animals going extinct so our children can see."

The relationships we formed over time with people from different villages created a special emotional bond. It was easy for them to be friendly toward us because our faces did not seem foreign to the Russians thanks to the ethnic Korean communities nearby; also, we didn't wear fancy clothes when we worked, and we spoke decent Russian.

However, we sometimes ran into villagers inside the nature reserve, like we did with Tanya's parents. As unauthorized persons can be heavily punished for entering a nature reserve, it was awkward when this happened. It wasn't as bad when the villagers in question were ginseng gatherers who didn't harm wild animals, but things became strained very quickly when we encountered a villager carrying a gun. Some of them would sneak into the nature reserve during the holidays or at the end of the year to hunt meat for seasonal feasts. The high price of pork and beef drove poor villagers into the woods, and poaching became increasingly worse around the holidays.

While living in the mountains, we often caught people in the act of poaching. Naturally, we came to know who caught what. We were obligated to report them to the authorities, but we couldn't bring ourselves to do it because we understood that they wanted something for the holidays and their children. I had once ran into a man from the village who had brought his wife out hunting. It was a strange encounter, but also very sad. Forest rangers would sometimes track down poachers only to find out they were friends. One side needed to eat and survive at any cost, and the other needed to protect at any cost. We couldn't treat them like criminals, but not simply because we were their friends. They weren't professional poachers. They would never dare dream of going after a tiger or any other beast that size. All they wanted was to shoot a deer or boar so they could have meat for dinner.

The real problem we faced was commercial, systematic poachers—professional hunters, mafia with capital, or the army with an arsenal. They targeted deer, boars, bears, and even tigers. Since it was nearly impossible to shoot a tiger in person, they installed booby traps and snares in large numbers. The booby traps were so well camouflaged that they posed a threat to forest rangers as well. A large tiger went for $20,000 to $30,000 on the Chinese market, so guns were a worthwhile investment.

After huge Asian markets dedicated to selling the parts of poached animals opened in the wake of the fall of the Soviet Union, a wave of Chinese merchants arrived in Ussuri. Ads for tiger sales and tigers wanted for purchase ran on TV and in newspapers. Tigers were sold whole or divided into parts—bones, skin, and meat. The majority of the suppliers were Russian mafia, military, and professional poachers, and the consumers were mostly Asian—some Koreans and Japanese, but mostly Chinese. The price was determined according to the size and characteristics of the tiger. Tigers bred in captivity went for $3,000 to $5,000. Wild tigers went for upwards of $30,000. For the Russian economy in those days, these were enormous sums.

Once, a man named Lumiere asked me for some footage of tigers. He was the boss of the second-largest crime organization in the eastern Russian city of Khabarovsk. When I asked him what the footage was for, he said it was for a TV ad he was going to run in China to spur sales. I was harassed quite a bit after I refused. The following year, Lumiere was caught in a struggle between crime organizations and was shot in the head in front of his house. The influence of the Russian mafia reached far into the forest. Mines, oil wells, lumber, and even the tiger skin business were often controlled by the mafia. If a member of the mafia became a state governor and started to lean in favor of one family over another, he could be assassinated.

In the past, the region north of Mount Baekdusan was called Suhae, or "sea of trees," and the region north of the Tumen River was called Millim, or "the jungle." The forest was so dense in this region that it was nicknamed the "Amazon of Northeast Asia."

The Ussuri forest, far more treacherous and sparsely populated than the forests of Manchuria, was especially well preserved. But since the migration of the Han people to Manchuria in the nineteenth century, the sea of trees and the jungle had all but disappeared and the trees along the Tumen River had been cut down to maintain an unobstructed view of the border and North Koreans attempting to flee. Trees several hundred years old were cut down in the wake of the Slavs' southward migration to Ussuri, leaving behind the smaller trees. Forests shrank, and animal populations decreased, taking the tiger population down with them. Today, people gun down the few tigers that remain. The forest village people shoot to survive, and the outsiders shoot for greed.

CHAPTER 19

White Snow and White Sky's Dispute

TO WAIT FOR a tiger is to wait for yourself. In my new bunker on the beach, I waited each day for a tiger that did not come. I turned on the camera every ten minutes in anticipation. The wind squeezed its way into the bunker to seek shelter from the cold, and the light avoided the bunker's darkness. Each time I switched on the camera, only the cylindrical haze from the viewfinder lit the inside of the bunker. A day went by. A month went by. I began to wonder. Had there been an accident? Had the tigers changed routes? As the days passed, anticipation turned into doubt—*They're not coming today, are they?*—and as the months passed, doubt turned into negativity—*Of course they're not coming today.* It was easy to fall into the trap of time and lose focus.

The days of waiting stretched on indefinitely. There is a mere twenty-four-hour space between the day a tiger comes and the day no tigers come. I had to keep believing that these two days were part of the same continuous flow, not separated by a boundary into black and white. I tried to concentrate by telling myself, *They haven't come by in two, three months, so it's likely they'll come soon.* Waiting for a tiger is like running a marathon. You must pace yourself and try not to push too hard. Just as the length of those forty-two kilometers is made up of little steps, I had to keep my pace and prepare for the final stretch.

If I divided my days into ones where I saw tigers and ones where nothing happened, I lost hope and fell into despair. But if I saw time as a continuous stretch, I was able to prepare myself for when tigers would show up. I anticipated every small possibility and made preparations, everything from ensuring I had a backup battery to adjusting the position of my water cup. I had missed opportunities in the past when a tiger came by because I was short just one 1.5-volt battery, or I had put my life in jeopardy because a water cup had fallen off a shelf. I now made minor adjustments each day and tried to tell myself that a tiger was coming the next. The longer they didn't show, the more I convinced myself their appearance was imminent.

At the end of these hours of anticipation and small preparations, a tiger showed up without warning or ceremony. When it emerged from a snowy shrub, bringing with it a chilly energy, I felt something warm rising inside of me. *You've returned alive and well. You are living life according to your own pace.* The relief penetrated through the long hours of waiting, and tears welled up in my eyes.

But the sentimental observation was brief. My sense of relief was quickly replaced by a live, pulsating tension. Like a marathoner on his last spurt, I felt as though I was about to breathe my last breath and every blood vessel in my body was about to explode. The brief moment when I filmed the wild tiger felt like an eternity. I was awash with ecstasy, but the moment didn't last. Like marathoners dispersing after a race, the tiger left soundlessly. When I woke up from this dream that sent my heart racing, I found myself alone at the starting line again. I pulled myself together by telling myself, *To see a tiger, you must see yourself and submit to nature.*

There had been no news of Bloody Mary's whereabouts for two months. Since her bunker attack at the end of the previous year, she had not been seen. She'd been unable to confirm my presence then, but she must have noticed something. The whereabouts of the three siblings were also a mystery. Since White Moon had gone inland and White Snow and White Sky had headed for the coastal ridge, we had found no trace of them.

I got a radio transmission from the Ural, our base camp on wheels. The tracks of two tigers had been found near Petrova Lodge. One was small, the other was large with the paws of a male tiger that was still growing. It was probably White Snow and White Sky. The sister and brother had followed the tracks of the four-eyed dog Chara to Petrova Lodge, turned around in front of the lodge, and headed for Bishannie Beach.

The siblings were fully grown, but they hadn't matured mentally. They lacked the prudence to stay away from humans. If they were with Bloody Mary, they wouldn't have gone near the lodge. Bloody Mary must have taught them to avoid manmade structures, but they hadn't experienced how truly dangerous they could be. It took surviving a great deal of hardship to become a true wild tiger. The reality of life in the forest is so trying that survival is difficult even for experienced wild animals.

Four days later, a bird picking at a rosegold pussy willow on the frozen brook flew off and shook the willow branch. A small pile of snow fell off to reveal a purple willow bud wrapped in layers of fuzzy scales. Spring had found the willow amid the brutal cold.

I found tiger prints next to the red stigma of the rosegold pussy willow. I zoomed in and focused the lens. One set of prints was small, the other large. The small prints were superimposed over the large ones. The big one led, and the small one followed. It was White Sky and White Snow, the siblings who had visited Petrova Lodge a few days earlier. The tracks followed the little brook down to the beach. Seeing those clean tracks set off a great ripple deep in my chest. The thought of the living creatures that had made these tracks caused my heart to beat wildly. I wanted to crawl out of the bunker and follow the path of the white flower-shaped prints.

I fought the urge and traced the tracks with my camera lens instead. Alongside the tracks was a long, shallow furrow, similar to the trail left by a rolling ball. The marks became more pronounced, and soon they were everywhere, as were haphazard tiger tracks. In the midst of all this lay a yellow ball. It was a plastic buoy.

White Snow and White Sky had played with a buoy that had swept ashore. I saw signs of tigers running, jumping, and frolicking with the ball. I couldn't figure out the whole sequence, but it looked like they had passed the ball back and forth and then White Sky had rolled and dribbled it. White Snow had tackled him to intercept, and White Sky had pounced on the ball to stop it from rolling back into the sea, then had kicked it back up to White Snow. In the middle of the night when the world was asleep, Bloody Mary's energetic children had played soccer here. I saw a childlike innocence, not yet tainted by the realities of the woods, in this scene of lighthearted play.

The next set of marks showed that the brother and sister had suddenly sprung up from where they'd been resting in the snow and charged forward. The marks ended at spatters of blood. There was a deep groove in the snow where the bloodied thing had been dragged. The groove continued to the rock cliff by the beach, at the bottom of which lay a dead doe. The artery in its neck was ruptured, and the area was soaked with blood. Its face was bloody and its jaw was shattered. The hunter hadn't just severed the artery—he or she had bitten down so hard that they had crushed the doe's jaw. I couldn't tell if it was White Snow or White Sky, but the cub had learned well from its mother.

The siblings must have spotted the deer while resting in the snow after their soccer match. The day before had been the full moon, which meant the thunderous sound of waves and eddying of the ocean winds would have masked the sound and scent of the tigers. The doe's senses must have been so overwhelmed and paralyzed by the forces of nature that she had failed to notice a tiger—two, no less. After the siblings had hunted the deer, they had dragged her to the bottom of the cliff, but had hardly eaten. They must have hunted at dawn when the tide was at its highest. They had taken a few bites, then headed back into the forest when the sun rose.

When a tiger hunts a deer, it eats it over the course of two or three days, divided into two or three meals. Each meal takes about an hour. On the first meal of the first day, they rip out the rectum and eat the intestines. They drink the blood and the warm liquid in the intestines. Next, they pull out the fur and begin to eat the rump where the meat is the tastiest. On the second meal of the first day, they eat the remainder of the rump. They flip the deer to maintain symmetry as they eat. We sometimes found deer carcasses in the woods with just one side eaten. Those deer had died of hunger and disease and then been nibbled at by the scavengers of the forest, such as crows, eagles, martens, and raccoons. To flip over a deer or a wild boar, a predator has to have the strength of a tiger, leopard, or, at the very least, lynx.

On the second day, the tiger eats the back and chest of the deer, and then whatever is left on the third day. There's not much remaining by then, so they lick the meat clean off the bones. If they're still hungry, they'll also eat the scalp and the stomach (but not its contents). When a tiger is eating, it concentrates on finishing off the prey it has caught instead of hunting new prey, even when there is prey nearby. This shows that the Siberian tiger is sensible, not greedy.

THE FULL MOON rose, and a dark shadow lurked on the coastal cliff, which was saturated with moonlight. The shadow seemed to be eating something, as it alternately lifted and lowered its head.

The beast cast a large shadow on the cliff. Another shadow grew as it approached. The crouching shadow suddenly jumped to its feet. The approaching shadow shrunk as it retreated. Each time the first shadow reclined and stood, the second shadow advanced and retreated. As they repeated this dance, the second tiger's shadow appeared to be flapping in the wind that whirled off the coastal cliff.

White Sky and White Snow were back for the deer. But something strange was going on. The affection of yesterday's play session was gone, and one sibling was shooing away the other. The one with the deer dragged it into the dark forest. The other followed. On the breaking waves, moonlight glittered like a school of silver fish.

The next day, I spotted a deer carcass in the young pine grove under the drooping green branches. A row of shrubs stood next to the half-eaten deer, and beyond that was the blue sea. This was the very spot Bloody Mary's family had shared a deer the previous summer when Khajain was visiting. In the night, the tiger had dragged the deer into the pine grove and eaten more than half of it. I kept the lens focused on the pine grove and turned on the camera every ten minutes.

Around four in the afternoon, White Sky appeared from the far side of the pine grove. I was surprised that he had shown up so early and wondered whether the gray sky threatening to snow or his careless immaturity was to blame. I hadn't noticed it the previous night, but he had grown in the two months since I'd last seen him. His ruff was fuller and his shoulder blades jutted out. He looked like a real Ussuri tiger now. In this period right before independence, male and female tigers show a stark physical difference in terms of bone structure and size. Males grow miraculously fast and develop bolder personalities.

I carefully focused the lens on White Sky. When filming a tiger, you must keep your eyes on their eyes to gauge their thoughts and determine whether it's a good idea to move the lens at any given moment. It's dangerous to film a tiger without keeping its eyes in the viewfinder. When I find myself unable to see the tiger's eyes while filming, I move the lens as little as possible, slowly zoom out, and find a moment when the tiger is not looking to fix the camera angle.

White Sky made sure that the area was safe before sitting down to eat his deer. Just then, White Snow appeared behind White Sky's wide back. She slowly advanced toward her brother, who sprang to his feet and growled at her. White Snow hesitated, and White Sky returned his focus to his deer. This was a very different picture from what I had seen two months earlier when they had gotten along. White Snow hesitantly inched forward again. White Sky roared even louder and lunged at White Snow. He couldn't bring himself to bite her, but stood with his face right in hers. He threatened her with a terrifying growl drawn from deep within that vibrated in his throat on its way out. His ghostly breath struck her brow. White Snow stared blankly back at him, not understanding what was happening. As her brother continued to growl, White Snow skulked back in retreat.

Now that he had chased White Snow far enough away, White Sky returned to his deer. Pretending to sniff the base of a pine, White Snow glanced over at White Sky. She continued to pretend she didn't care as she slowly approached again. Without turning around, White Sky let out a low growl as he ate his deer, as if to say he knew what she was up to. White Snow, still hesitant, plopped down on the ground and watched him eat. White Sky didn't care if his sister was watching or not; he finished off the rest of the deer.

They were no longer affectionate siblings who shared what their mother hunted for them. White Sky had entered a phase in his life in which he felt more comfortable being alone. He was going through the same rite of passage that his father and his father's father had gone through—one that everyone goes through.

Like spring acorns budding into oak sprouts competing with one another for soil and nutrients, the brother and sister would fight each other for larger hunting grounds before long. When he was older, White Sky would have to compete with the alpha male of the region, Khajain, the Great King and his own father. Khajain had gone through the same thing, as had Tail and Kuchi Mapa, the two Great Kings before him.

After brother and sister left, snow fell on the empty forest, piling layer by layer on the young pines. Time passed in the bunker, too.

CHAPTER 20

Season of Melting Snow

I STARED OUT OF the bunker all morning. In the warmth of the sun, I could almost hear the ice cave of the northerly winds closing up. The peaks and mountainsides were all covered in snow, but in a sunny spot at the bottom of the hill, sunlight glistened on dry blades of grass. Yellow Amur Adonises had sprung up in the snow. Drawing from Mother Nature's energy, it was once again the first flower of the year to bloom.

The more I looked at the sunlight, the sicker I felt inside. I crawled out of the dark bunker and dragged myself to the sunny spot at the bottom of the hill. Was this how bears felt when they came out of hibernation? Sunlight split into rays and spun overhead. I was shriveled, weak, dizzy. I squatted on the ground and soaked up the sunlight. The rays felt warm on my skin.

On the other side of the hill, a wildcat was also getting some sun. It looked unwell with its unfocused eyes and chunks of missing fur. Too weak to muster up the strength and resolve to run away, it turned its haggard face and gazed at me.

A cold wind swept by. A small whirlwind perked up the limp, dry blades of grass. Bits of the wildcat's fur fell out and blew in the wind. Greasy clumps of my own hair fluttered in my emaciated face. Dandruff fell from my dirty, smelly, dry head. The flakes landed on a wormwood bud wrapped in fuzzy scales. The wormwood had been quite green the previous summer, but had dried as autumn passed, and the winter winds had snapped off its brown stigma. Nature's skin had dried and shriveled like mine as it had soundlessly withered, spent the winter under snow, and then revealed itself again. Life was blossoming again under the layers of dead skin nature had shed through the winter. Tears flowed from my wasted body. Like the wormwood and the wildcat, I had lived life head-on. Time, circulation, cycles—their futility and eternity were stark and unavoidable in the sunlight. Perhaps spending half a year observing nature in one spot was a sad thing.

The days were still cold, but all living things know there are always warmer days to come. There are lives that cannot hold on despite this knowledge, and those that are full of life because of this promise. The spring draws a clear line between those on their way out and those on their way in. The molting wildcat staggering away reminded me of myself. I wanted the droning waves to give it a rest for once.

I received a radio transmission. The villagers at Mayak Village had reported seeing a tiger limping with an injured paw around Mayak Lake. The sound traveling through the radio waves was as hoarse as the echoes in a humid valley. I thought of the military base in Mayak. There had been a rumor going around since the previous summer that the army planned to install booby traps all over the forest. At the same time, I thought of White Snow and White Sky playing soccer with a buoy on the beach. They were the only two tigers we'd seen lately in the coastal area.

I decided I must leave the bunker. I stuffed a few simple camping tools in a small knapsack and climbed over the range. My muscles weak from lack of use, I was quickly out of breath. My knees kept buckling. After three mountains, I saw Mayak Lake. The cold air of early spring had met the humidity of the lake to conjure up a thick, gloomy fog that hugged the mountain's waist.

I saw a small crowd of people standing in an oak grove on a slope. On this north side of the mountain, the snow was still as dry as frost. On that white snow, Bloody Mary lay in a pool of blood so big and so red it resembled spilled paint. Her eyes wide open and her jaws clamped shut, she was stiff from head to toe. The leaves on the young oaks that surrounded her trembled in the wind. I couldn't hear anything, and my mind went blank. I had a stabbing headache.

Bloody Mary's left front paw and shoulder were shredded. That seemed to be the source of the pool of blood. She'd been hit by a shotgun blast. But the real fatal wound was the one in her side. A bullet, probably fired by a Russian army rifle, had entered through her lower right side and exited out her left. The entry wound was only the size of a finger, but the exit wound was as big as a fist. Blood and intestines had spilled out of the hole, turning everything around them red.

The villagers—Tanya among them—explained. The previous morning, a few villagers had been fishing at Mayak Lake, which had begun to thaw. Tanya had been watching them fish. Wild ducks had been swimming in the icy water for days and bobbing among the chunks of ice. Tanya watched the ducks move as quietly as the mountain shadow on the water. Suddenly, the ducks kicked the calm surface of the water and squawked as they flew up. Turning to watch the ducks circle over the lake, Tanya had spotted the tiger. It was slowly limping down to the edge of the water.

"Amba! Amba's here!" she'd shouted.

The villagers turned and saw the tiger. When the crowd began to murmur, the tiger roared and charged at them. But she couldn't run very well. She was dragging her front paw, and something was flapping on her side. When everyone fled, the tiger stopped. She stood by the lake for a while and then returned to the forest. Running away, Tanya continued to glance over her shoulder at the tiger.

Bloody Mary, who had spent her entire life avoiding people, had willingly come out of the forest and walked along the edge of the lake. She couldn't put weight on her left front paw. Even in her condition, she charged at the fishermen for thirty meters, and she was so filled with rage that she was able to do it with her guts spilling out. She did not pursue them any farther when the people dispersed. She returned to the forest and walked for about three kilometers. Her tracks drew an irregular zigzag. She'd stopped to rest several times. She walked through the forest off the beaten path, but was headed for Diplyak. Bloody Mary had lain down for the last time on the last slope to Diplyak and had continued to bleed. I think she died from blood loss.

Before Bloody Mary had come down to the lake, two villagers (named Lee and Misha) had seen her. They were hunting around the lakes for wild ducks that had begun their migration the week before. Lee was an ethnic Korean living in Russia, but he did not speak Korean.

"We were looking for wild ducks near the lake," said Lee. "We were on the forest path when we saw a tiger coming down, over there. We quickly climbed oak trees to get away. When we looked down at it, we saw that the tiger was staggering a lot. Guts were spilling out of its sides, and it was very bloody. It came all the way to the bottom of the tree, growled up at us, and continued down to the lake."

I went to the spot where Lee said he had seen Bloody Mary. The narrow mountain path he mentioned circled the lake and led up and down a steep slope. It was a well-beaten path used by many animals and the humans who pursued them.

Lee and Misha had been walking around the lake armed with an old shotgun loaded with bullets the thickness of several pieces of classroom chalk and some military-issue binoculars. Lee had been slowly approaching the ducks quacking and flapping just above the water when he had seen the big tiger coming down the slope opposite him.

The traces Bloody Mary left on the slope revealed that she had been bleeding, but not limping. Both front paws had left clear tracks on the snow. Then the tracks showed that she'd begun to run. Bloody Mary had run all the way up to the oak that Lee and Misha had climbed. There was mud on the branches and there were claw marks on the trunk. Bloody Mary had scratched the bark as she jumped. There were tracks of her landing after the first jump, and then jumping sideways. From there, the tracks led down the mountain, her left paw dragging in the snow. I found two yellow plastic shotgun shells, one near the oak and the other on the mountain path. Lee was lying.

Here's what had really happened: Lee had seen Bloody Mary coming down the slope. He'd seen that the tiger was injured and dying, so he'd shot her. She'd charged at him. Lee and Misha had climbed the tree. When the tiger continued to threaten them, he'd fired again from up in the tree. Bloody Mary had been hit in the left shoulder and paw from a close range. When the shot was fired, she'd instinctively jumped sideways and limped down the mountain to get away from the gun. Shot through the stomach and already at death's door, she now had a new shotgun wound. Bloody Mary had lost her composure. When she saw people by the lake pointing at her and murmuring, she'd charged at them, her rage flaring up anew.

Bloody Mary had come down along the mountain range from inland before running into Lee. I found blood trails all along the range. I followed the blood up around four kilometers to find an old birch at the end of its life. Tiger fur and claw marks were on the trunk, and the stench of tiger urine was strong. This was a marker tree for tigers in this area.

At the base of the birch, dry branches and leaves had been raked together into a mound and then hastily scattered. A booby trap was installed there. The camouflage branches remained, but whoever installed the trap must have come by to collect the gun. A few sets of army boot tracks skulked in the area for a while before disappearing down the mountain. About one and a half meters in front of where the gun had been, a tiger track was trampled on by a boot track. Bloody Mary had been shot here on her way down the mountain. She had sprung up the moment the bullet had hit her in the side and spilled her intestines where she landed, leaving a lump of frozen blood.

Bloody Mary carefully put one paw ahead of the other. It had been two and a half months since her visit to the coastal range. She had been busy patrolling Deer Valley and Crow Mountain, resting in the pine nut grove, hunting a wild boar, and stopping by Dipiko and Santago to mark her territory. She had also traveled down the Tachinko coastal cliff where the mountain goats live and marked her territory on Yaloan Tuke. According to the traces on the old tree, Khajain hadn't returned from his pilgrimage north.

Bloody Mary worried about her cubs as she headed south along the coastal range. White Moon had grown accustomed to being alone and was a skilled hunter, but White Sky and White Snow were having trouble becoming independent. They were still together all the time, even though they should have gone their separate ways by now, and they were still following Bloody Mary around. If she hadn't killed that wild boar for them in the pine nut grove and quickly left them, they would be trailing her right now. As soon as they were independent, she'd take some time for herself and prepare for what might be her last litter.

As she got closer to Diplyak, she remembered an incident from long ago. She'd picked up a strong scent of human. She thought about going down there again to check, but talked herself out of it. She was just going to follow the mountain range, see what the humans were up to in the village, and head back inland. Spring was near, which meant the badgers and bears would be coming out of hibernation soon. She made a mental note of the few badger and bear dens she had found at a sunny spot on the Dragon Spine.

The lake was just beyond this hill. Heading down the mountain range, Bloody Mary was as excited as if she'd already killed a bear. She saw the old birch where she usually marked her territory when she was in the area. She hastened to the tree to see if her cubs had left their marks there.

But the closer she got to the birch, the more she felt something was off. She looked around, but saw no sign of humans. There was a faint but sure smell of metal. She caught a whiff of her cubs' scent in there as well. She took another step toward the birch to get a better whiff. Her paw caught on something. There was a loud bang, and something hit her side. Bloody Mary jumped. Hot blood poured out of her. She saw the gun barrel hidden at the base of the birch. She was consumed with rage. She went around the tree, dug the gun out of its camouflage, and broke it apart. But she knew it was already too late. The wound was too deep. Realizing what was about to happen, she continued down the range.

A crow spotted the carcass and flew over. It sat on a branch above Bloody Mary's body and craned its neck.

"Shoo!" Stefanovich threw a stick at the crow. "There's nothing for you here!" The crow cawed a few times and flew away. In the far corner of the sky, eagles, small as black dots, dove and swooped right back up into the sky, which stretched on infinitely. They drew great, overlapping circles in the sky, as free as could be.

Spent and weak, Bloody Mary had lain here in the snow. She had kicked the snow with her hind legs and scraped it with her front paws. She must have been kicking right up until her very last breath. The struggle and pain of her last moment were apparent in her wide-open eyes. It wasn't the kind of death that nature brings like a gift of eternal rest at the end of a fruitful life. I stroked her head and whiskers. Her whiskers were stiff. I remembered how they'd felt on the back of my hand. I closed her eyes.

Tiger tracks circled Bloody Mary's body. One set was small, the other large. It wasn't clear if they had been there when she died or if they'd found her body later, but brother and sister had stayed by their mother's side for a long while. The surrounding area was full of their tracks and there were traces of the two lying in the snow waiting for their mother to wake up.

The tracks circling Bloody Mary finally departed. The small paw prints were sometimes superimposed on the large ones. White Sky led and White Snow followed. Their tracks headed straight up the range, then paused on the mountainside. White Snow had lain in the snow like a sphinx, facing her mother on the slope. White Sky's prints circled the sphinx, perhaps urging her to hurry. They had set out again. The tracks disappeared over the range. The siblings had bid their mother goodbye.

PART IV

New Generation

CHAPTER 21

Commotion in the Forest

THE NARROW DEER River folded and unfolded on its way into the endless sea of trees. At the tail end of a fading winter, the river was lined with towering oaks on both sides, and the ones that had fallen blocked my way as I hiked along. A soft sheen of moisture coated the water's surface, which had been frozen solid all winter. I held my ear just above the surface and heard water coursing underneath.

In early March, I moved to the bunker in Deer Valley. I tidied up the bunker and placed the camera on the tripod. Inland bunkers had a different feel than the ones on the coast. The air was cold but fresh, and I heard the faint sound of birds chirping and a woodpecker pecking. The naked forest was full of stillness rather than the raucous sound of waves. The stillness would persist until the ice thawed in Deer River and the water began to babble.

White Moon had not shown herself on the east coast for over three months, but tracks of a lone female tiger had been found all over the Sikhote-Alin Range, from Black Mountain and Crow Mountain to Deer Valley in the south. It seemed White Moon had claimed an inland territory.

Tigers are conscientious animals. They are constantly on the move, traveling over ridges and around rivers. Except under special circumstances, however, tigers do not move hastily. They travel steadily and carefully with the morning dew on their backs. In the early winter or early spring when there's less snow, they stick to the frozen rivers. The rivers in mountain valleys form thick ice that makes a good route for animals even after the snow has completely melted away. At that time, the deer begin to return to Deer Valley after spending the snowy winter in the mountains or on the coasts.

When tigers move, jays, crows, and eagles move with them, hoping for the remainder of the tigers' kills. Tigers also visit places where crows and eagles have congregated to see what the forest chatterboxes are up to. They meticulously keep track of everything that goes on in their territories, such as intruding tigers and hunters, and note the changes since their last visit. This sensitivity to changes in their territory allows tigers to avoid danger and competition. So whenever there's a commotion in the forest, tigers investigate. The Udege refer to this tiger behavior as "reigning over the forest." To the Udege, the tiger is the true lord of the forest.

A BLACK WOODPECKER adorned with a red cap pecked at a willow on the other side of the river. There was a hole in the willow about five centimeters wide and a small mound of wood dust below. This would be a new home for this year's baby birds. Behind the willow was a sunny hill with a raccoon nest. The raccoon sauntered up and down the river looking for food on warm days and returned home when it was full.

One morning, as the sun beat down on the snow, a deer carcass was discovered across the river from the raccoon nest. Excitement shook the stillness of the forest. A wood nuthatch, whose delicate cries always made the calm of the morning even calmer, was the first to fly over and peck on the deer for breakfast. A great spotted woodpecker joined the wood nuthatch and pecked pretty vigorously at the deer, making a dull sound with each jab. A yellow-throated marten heard this and joined them. Instead of nibbling at one side, the golden-furred animal jumped from one side to the other looking for a tender spot.

A crow perching on top of a tree cawed. *Marten! Marten eating deer!* This prompted a flock of crows to fly over and make a big fuss. The marten was busy eating and shooing crows at the same time. Shoo one crow, another crow lands. Shoo the other crow, and the first crow comes back.

Just then, from on high, a vulture circled overhead and dove at the deer. The vulture puffed up its neck feathers and screeched at the scavengers. After a few halfhearted attempts to fight, the marten scurried off. More vultures arrived; the flock grew from two, to three, to a dozen. They spent more time pecking at each other than at the deer. One climbed on top of another and jabbed ferociously, and another joined in the fight. Yet another ambushed its fellow birds with momentum as it descended from above. One of the weaker vultures that had not gotten enough food in the winter died in this pandemonium. The other vultures fed on the dead one. The crows took a bite of deer here and there while the vultures were occupied with fighting each other and flew off when the vultures came after them.

The raccoon came outside to see what the birds were making such a fuss about so early in the morning. It stood there watching the vultures, but didn't dare get closer to the deer for a morsel. Also pushed to the sidelines by the vultures, the crows started to pick on the raccoon. One hopped around the raccoon to intimidate it, while another hovered over it and nipped at it when it could. A third one sat on top of the raccoon, and when the raccoon snapped at it, it flew off.

A few golden eagles circling the sky flew over to see what the fuss was about. They waited and watched the vultures' rumpus.

An otter peered out of a hole by the river with snow on its head, and a young lynx emerged from under a snowy log. Its sleepy face gazed about as if to say, *What's going on?*

Springtime in the forest is a time of poverty. The remains of those that did not make it through the winter appear, and the surviving but starved fight each other to the death for those remains. New casualties result from this process. The quiet of the spring forest is instantly disturbed and the impact travels far.

The black woodpecker was indifferent to the racket. It pecked away in a peculiar rhythm that reverberated through the forest. Suddenly, the pecking stopped. The black woodpecker stared intently at one corner of the forest. A familiar shade of orange lurked in the thick woods on the other side of the river. The creature came over the hill in great strides and stretched straight up against a large tree.

Caw! Caw! Caw!

The crows shrieked urgently and took off in unison. The sky was dizzy with the sound of cawing and flapping. The raccoon seized the opportunity and darted at the deer.

The tiger carefully studied the river and then looked down at the raccoon with its tense, fierce eyes. It was White Moon. Her bone structure had widened and she had put on some weight. She was taller and her fur was well groomed—she had grown into a mature, independent, beautiful tigress. She must have dropped by on her way back from patrolling her territory along the Sikhote-Alin Range.

White Moon waited for the vultures and crows to settle down. Mere months ago, she would have snarled and attacked a pestering crow, but she had become much more careful.

She headed along the river toward the raccoon. The smaller animal belatedly saw her and scurried downstream. Enervated after being cooped up in its den for most of the winter, its legs did not cooperate. It stumbled and fell several times before it was safely out of sight. But White Moon didn't so much as glance at the raccoon scrambling for its life. She cautiously examined where the vultures and crows were fighting each other and gave the deer a thorough sniff. She looked up and down the river again. Hunting was not her objective. She was here to find out what the racket was about.

White Moon sniffed at a pile of logs on the riverbank. The trees had been swept downriver in the flood the previous summer. We had hidden a wireless microphone in there to record the sounds of nature.

White Moon snorted, a puff of breath pluming from her nose, and knocked over the pile of logs. Through my headset, I heard the thunderous sound of snapping wood. White Moon had found the exact location of the tiny *plastic* microphone. She ripped it out and chewed it to bits with her incisors. She then circled the area for other strange scents.

Keeping track of any commotion within its territory and checking for unfamiliar smells is a sign of a tiger's affection for its home. Like a lord in the Middle Ages, White Moon understood that Deer Valley was her land and must be looked after. This was the behavior of a mature tiger.

After a thorough investigation, White Moon waded into the middle of Deer River. The sun was setting over the lower end of the river. She gazed downstream at the iridescence. She stood with her back to me, facing the river and the sunset. She reminded me of a cowboy in the Wild West, tension beating just below the calm surface. Was this what it meant to be a wild tiger? White Moon's orange, glossy winter coat looked fiery in the evening sun. It was like seeing Bloody Mary again. Did White Moon know that her mother was dead?

Now more relaxed, White Moon took a sip of water from a puddle of melted snow. She gracefully shook the water off her front paws and sat down by the water to rest. She liked the secluded river deep in the forest. She licked her paws and groomed her fur. She yawned, revealing her large incisors. She rested now with her eyes closed. She seemed tired but comfortable. She took a good thirty-minute rest without once glancing about her.

White Moon got up as the last light of the day disappeared over the forest. She walked along Deer River. For an independent tiger, constant surveillance of its territory is just as important as hunting. Once again, White Moon set out on a long journey along the Sikhote-Alin. Blue evening light outlined her silhouette as she walked. Her slow saunter was partially obscured by the branches hanging over Deer River, and eventually she disappeared.

CHAPTER 22

Tigers and Azaleas

I SCALED AZALEA CLIFF, where shimmers of heat rose. The higher I climbed, the steeper the cliff became. The azaleas had already bloomed along the ridge and on the dizzying cliffside. Some had not yet emerged from their plump buds, while others were in full bloom. The light pink petals fluttered in the wind.

From the top of Azalea Cliff, Triparashonka Beach came into view. Ussuri tigers watch over their territories from spots like this. There were tiger traces on the base of the pine at the cliff's edge. After Bloody Mary died, White Snow and White Sky had traveled along the coastal cliffs, looking for ungulates in the coastal basins. Among these, Tachinko Cliff was the most dangerous, followed by Azalea Cliff.

I crawled into a bunker dug into a ridge that jutted out toward the coast; from there, I had a clear view of Azalea Cliff. The area was full of rocks and the bunker was right under a pine grove, so it was smaller and more crowded than an average bunker. Moisture rose up from the bunker floor, making it humid inside. I aired it out, finished preparing for the stakeout, and rebuilt the camouflage with leaves and fallen branches.

It was April. We were behind schedule. It had been over six months since I'd started the stakeout the previous October. We should have been done by now, but Bloody Mary's death had set us back. I wanted to make sure White Snow and White Sky were faring well after their mother's death, and I wanted to get a shot of a tiger standing over a cliff with azaleas in full bloom. We decided we'd stay until early May, whether the tigers came or not. The efficacy of the bunker declines dramatically after May. Lush vegetation makes it difficult to see if a tiger is approaching until it is right in front of the camera, and the rising temperature leads to a stronger human stench that increases the likelihood of being found. In the winter when the winds threaten to penetrate your skin, all you want to do is stay in your burrow. But as the spring days grow warm, the urge to be out in the sun becomes hard to contain. At this time of year, the mental toll of being trapped in an underground bunker is especially excruciating.

I detected movement on the cliff in front of the bunker. I slowly turned the lens. I zoomed, found the focus, and locked the frame. All I could see in the dusk light were the contours of the terrain. Then I saw a shadow bobbing above the darkened cliff. It looked like an animal. From the way its head was moving up and down, I assumed it was feeding on something. The head came up again. It had pointy ears like a cat. Was it a wildcat? The head bobbed up again. The tips of its ears were furry and black like the tip of a brush. It was a lynx. It must have caught a turtledove on its way to bed.

Along with the leopard and the tiger, the Siberian lynx is one of the three large cats of Northeast Asia. It's the smallest of the three, weighing a maximum of about thirty kilograms, but also the fiercest. An Udege hunter named Tipui once saw a lynx kill a Manchurian red deer ten times its size. He had told me the story.

One autumn, during the Manchurian red deer mating season, Tipui had been hiding near a deer trail when a medium-sized lynx came along. The lynx took a running start, sprung off the ground, and clung to the trunk of a Siberian dwarf pine with its sharp claws. It climbed the tree, scratching the bark with its claws on the way up, and hid itself among the tree branches that extended above the deer trail. After a long wait, a herd of Manchurian red deer, each about the size of a horse, came along. The lynx jumped down onto the back of a deer. The deer jerked about, trying to get the lynx off its back. But the lynx had a firm grip on its prey, using its sharp claws to dig deep into the flesh, and no amount of jerking or kicking made the slightest difference. The lynx rode the deer like a rodeo cowboy, and when the exhausted deer paused for a second, the lynx sunk its fangs into the deer's neck. It clung to the deer's throat for thirty minutes.

In the end, the deer gave in, even though it was nearly ten times bigger than the lynx. When the deer hit the ground, the lynx clamped down on the throat and shook even harder to finish off the job. Only when the deer was completely still did the lynx let go and raise its head. The gray fur on its intense face was stained red. Tipui said the lynx must have been exhausted, too, because it took a long nap afterward, using the deer as a pillow.

Even in this darkness, animals stood at the fork between life and death. The dark shadow of the wild animal on the precipice was a contrast against the dusky sky. Stars came out one by one. Over the cliff where the lynx stood, the moon stretched its slender limbs as it floated up in the sky. The pale moonlight shone on the melancholy coastal cliff. The forest was frighteningly silent. Even the waves were quiet, as if the ocean didn't want to be found. The wind alone whooshed past once in a while. In the stillness of the night, I heard the gentle breathing of the sleeping mountains. Sikhote-Alin was in a deep, deep slumber. A fox barked raspily in the distance.

The dark silhouettes of the pines standing on the cliff shuddered as the wind blew through them. The screeching of the fox piercing through the night sounded like the screams of the pines. Or the shouts of Kalgama. When the Udege hear an unidentifiable sound in the forest at night, they believe it's Kalgama.

Kalgama is a tree spirit that feeds on resin. His body is shaped like a tree trunk, and his head is like the batting end of a baseball bat. He is about half as tall as a pine tree and has two fingers on each hand like pincers. He lives on piles of old trees or firewood and wears a belt around his waist. Legend says if a brave man fights Kalgama and takes his belt, he will have great success in that year's hunting and become rich. Kalgama, by contrast, loses his power. So Kalgama wanders the forest at night, shouting, "My belt... Give me back my belt!"

If someone gets tired of hearing Kalgama's cries and so much as mumbles to himself, "Here, take the damn belt," the person will become poor. So when an Udege hears Kalgama coming, they avoid him to prevent their good fortune from being robbed from them and then return when Kalgama is gone.

There are other such relationships in the Ussuri forest—not one of predator and prey, but two parties that mutually avoid each other just for the sake of convenience. Manchurian red deer and Ussuri sika deer are one such pair, and the Ussuri tiger and Amur leopard another. Like an Udege avoiding Kalgama, leopards avoid tigers. Thanks to its sense of smell and the tracks tigers leave, a leopard can find out relatively quickly if a tiger is in its territory. A leopard will leave its territory when it realizes a tiger is nearby and return when the tiger is gone. They compete with each other for ungulates and land, and every once in a while, a tiger hunts a leopard.

About thirty kilometers north of Tumen River is the Kedrovaya Pad Nature Reserve, which was established specifically to protect Amur leopards. *Kedr* means "nut pines," and *pad* means "valley." Put together, *kedrovaya pad* means "the valley of nut pines." True to the name, a dense grove of nut pines does line the valley. Two Amur leopards live in the nature reserve. The female leopard had been born through three generations of inbreeding. A male leopard that comes by every once in a while has been breeding with his biological offspring for three generations. As a result, one of his female offspring died young, and the surviving female doesn't have the sharp sensitivity required to be a wild animal; her territory is small, she's not a very good hunter, and she doesn't react quickly when humans approach. On the upside, she's easy to capture on film. But even a leopard with such poor reflexes knows to clear out when a tiger is near. The inbreeding may have dulled her wild genes, but she still instinctively knows that tigers are trouble.

SPRING SHOWERS CAME. The rain soaked through the earth, which released its heat in the form of fog. The sea's moisture wafted over, thickened the fog, and cloaked the coastal cliff. The fog shifted each time the wind pestered, and the pines on the cliff revealed themselves through the moving haze only to disappear again. The pitter-patter of the spring shower fell on it all. The fog descended like a ball of wet cotton as night fell. In the pitch dark, all I could see was the pale glimmer of a ship heading into the ocean.

I woke at dawn to the sound of birds chirping. I opened the bunker entrance, and blue fog rushed in. It had become thicker in the night and it was now difficult to see the outline of the tree branches right before me. Birds chattered in the mist.

The birds flapped back and forth. Judging by the volume and coarseness of their chirping, I guessed, correctly, that they were Ussuri jays. The jays emerged as the fog slowly rolled away. Dozens of them were hanging from bushes, feasting on dry fruit. Morning dew splashed each time the jays landed on a bush. If they flapped in an effort to pry off a dry fruit, the sweeping of their wings blew the fog away. A halo appeared around the sun, and it looked like the fog was thinning. All at once, the fog evaporated and the skies were clear. The warm sun dried the wet earth. Every drop of water hanging from a branch or bush was iridescent. The jays had their fill of breakfast and returned to the forest one after the other.

The sun came up, then went down, then came back up again. One day, just another day in the endless cycle of night and day, the wind began to blow. Like a tiger clamping down and shaking a deer's throat, the wind bit and clawed at the forest. It swam through the trees by night and down to the beach at dawn. It swept across the beach, working up a dusty blue storm, and the ocean took a deep breath to hold back its anger.

New tracks appeared where the wind had been. Living, breathing life could be found at the end of these traces. On the sandy beach as clean as the beginning of time, a single line of flower-shaped paw prints appeared. At the far end of this line was the coastal range leading to Azalea Cliff, poking out through the clouds. The sun emerged and the wet, dark rocks on the beach sparkled.

Black stripes on orange fur flashed between the green coniferous trees and the pale pink azaleas. The tiger made its way down the cliff and stood at its edge, looking nonchalantly out at the East Sea. The branches of the coniferous trees lining the ridge shook each time the cold wind wrapped its fingers around them as it passed. The tiger gazed down at the pebble beach through the fine, trembling branches. It turned to look at the middle of the cliff, too. It saw the bear den it used to visit with its mother. The weather was warmer now. The bear probably wasn't there anymore.

White Sky had grown thin. All of his baby fat was gone. But his bone structure was wider, his face was more mature, and there was a depth and calmness in the way he looked down at the East Sea. Like a young man who had overcome the trials of his childhood, he had grown up to be a strapping tiger. Under the beautiful pines, he took a step closer to the edge. White Sky stood at the end of the territory he had inherited from his mother. He was patrolling his land. Beyond White Sky's thin face, the sharp lines of cliffs and rocky peaks jutted out like a rooster's comb. Green pines stood along the cliffs, and azaleas grew in the cracks of every rock. White Sky looked good here.

A black bird floated up with the ocean wind under its wings over the blade-like rocks, and then over the pines. White Sky turned as his gaze followed the bird. On the steep ridge, a wild bush was bearing navy-blue buds. Azaleas bloomed everywhere, just past their climax. There were hardly any fuzzy buds left. White Sky climbed past the azaleas and up the hill, his back arching and expanding fluidly. With the light hitting the ridge from the other side, the skyline was a clean contrast of black mountain and white sky, with shimmers of spring heat rising in between. The tiger walked along the ridge. His silhouette blurred with each ripple of the haze as he moved through the coniferous trees. He disappeared beyond the white skyline.

The day after White Sky's visit, White Snow arrived at sunset. She appeared on the forest path along the coastal cliff, the dark red ocean behind her. She looked rather more disheartened than nonchalant. She was even thinner than White Sky, but bigger than she used to be, and her face had the look of a mature tigress now.

White Snow stopped at the tree that White Sky had sprayed. She sniffed it, raised her tail high, and sprayed it herself. White Snow was trailing behind her brother. Her brother had grown accustomed to his independence, but White Snow was still torn between her wish to be with him and her need for independence. She followed her brother's scent because the world was still an unfamiliar place, but the time for independence was coming. She stopped at every spot where we had hidden a small sensor.

Her ability to detect anomalies in the forest had grown sharper. Noticing unusual changes and caring about her territory were sure signs that she would be independent soon.

After Bloody Mary's death, White Snow and White Sky had managed to make the coastal range their home. White Moon had claimed a home inland. White Moon and White Sky were independent, and White Snow would be too, before long. They no longer had their mother's care, but they were each going their separate ways according to what their mother had taught them in the forest they'd inherited from her.

White Snow climbed the ridge, which was drenched in the evening light. Dusk, like glowing red metal, slowly cooled and disappeared. White Snow vanished over the same ridge as White Sky. The empty skyline felt lonely.

I decided to conclude the year's stakeout. My hair was gray, my wrinkles looked like fishing nets, the bones under my skin cast sharp shadows on my face, and my eyes had sunk into my skull. At the thought of seven months of stakeout life coming to a close, I was hit with a fresh awareness of solitude that I'd been repressing, but even in the midst of this sadness, I thought of the summer research expedition and the next winter stakeout. At the end of one journey, I planned the next. I wanted to cling to a tree and weep. I burst into tears. Was I afraid of the dark woods? Or of my own desires?

The forest teeming with untamed life dissipated into the background, and I hungered for the civilized world. I would soon return there to recharge, and when I came back to the forest, it would feel like it had been a dream. I would find myself yearning for another new adventure.

CHAPTER 23

Chance Encounter

THE EAGLE CHICK whipped its head around and looked at me. Our eyes met. There was a bit of surprise in the way it stared at me, but it didn't look afraid. Its eyes seemed to ask, *What is that?* A fly wandered into its line of sight, and its eyes followed the new arrival. Flies swarmed the fish it was eating, and it lost interest in me. Consumed with the task of shooing the flies, it swung its neck back and forth. When a cool wind swept over it, the white-tailed eagle seemed to have forgotten my existence as it gaily flapped its wings, ready to fly off, seemingly with its nest in tow.

The young are innocent. They act foolishly, but they're not jaded. The eagle chick had been born that year, but already it was as big as its mother. It was so large that it filled the one-and-a-half-meter-wide nest in the oak tree. Its tail had turned white. I didn't know if it was a male or female, but it was the mother eagle's only offspring, brought up with care and love.

I was regaining my strength and rebuilding the muscles I had lost in the bunker. Stefanovich and I were on our way across the inland region, headed for the coast. We walked for a few days, researching the forests on both sides of Santago River by following a zigzagging path from one side of the river to the other. The buds were unfurling into leaves and flowers, and the forest was clamoring with life. The green grew greener each day and life streamed forth in every valley. An eagle flew overhead and a wild sika deer played in the clearing. Summer had come to the vast land of the tiger beyond the Sikhote-Alin Range.

We carefully ventured down to the river. An azure-winged magpie was pecking at a fish. The bird's head was black, its chest white, and the sky blue of its wings extended down to its long tail. Its colors were so brilliant it could be from another world. It bobbed its head and continued eating the fish. Just then, a handful of crows arrived and attacked the magpie, which had no choice but to retreat to a branch on top of a shrubby bush clover, where it helplessly waved its tail feather. The crows devoured the fish as if possessed.

Perched on a dead, dry branch hanging over the river, a white-tailed eagle was so still it seemed to be dozing. It watched the crows for a moment, suddenly catapulted into the air, and swooped down on the crows like a paraglider. Before its branch perch had time to stop shaking, the fish was the eagle's. The crows hopped around and cawed in protest, but a few pecks from the eagle were enough to keep them at bay for good. A few disgruntled crows pestered the magpie as though this was its fault. The magpie weaved through the rosegold pussy willows to get away from them.

Meanwhile, the eagle swallowed the fish whole. It must have still been hungry, for it looked around and then jumped into the shallows with its talons up. It flapped its wings, making big splashes, but to no avail. It finally gave up and landed on a rock, where it shook the water off and combed its feathers with its yellow beak. Then it lifted its tail feathers, left a splotch of guano on the rock, and flew upstream along the water. The blue Santago River wrapped around the thick forest and curved left with the eagle's path.

The chill creeping over my shoulders woke me up. The morning fog was meandering its way into the river and forest. It became thicker until it turned into a drizzle. A silence as deep as death traveled down the river that flowed as steadily as a pulse. I heard something approaching through the fog on the other side of the river. I heard hooves kicking pebbles into the water. A small group of Ussuri sika deer emerged from the fog and crossed the river. The sound of deer wading in water with their fawns broke the morning silence like a melody.

We found a shallow area and waded into the river, too. The current was stronger than I'd thought, so instead of going against the current, we crossed the river diagonally, moving a little downstream with every step. Spiders had spun webs on every rosegold pussy willow and baby rose bush. The masters of these webs were nowhere to be found; the only thing clinging to the webs was morning dew. Some webs spun by giant spiders were more than two meters wide, while wasp spiderwebs were finely spun and strong. The small and extremely fine webs of spiders I wasn't familiar with were hardly visible. I kept feeling them cling to my face.

The forest grew warmer by the day. Summer was now in its prime. The farther upstream we traveled, the more we spotted animal traces left by Asian black bears, wildcats, otters, and badgers. Even thrushes and moorhens, rare in Ussuri, made an appearance. Along the river, young zelkova trees, shrubby bush clovers, and baby rose bushes formed a grove, with Mongolian oaks, hazels, alders, and birches standing over them.

At a stretch of the river that flowed past a nut pine grove, we took off our backpacks and sat on a flat rock by the path to take a break. We had a simple meal of bread and sausage and drank from the brook. Stefanovich lay down on the rock and started to snore. I sat, all strength drained from my limbs, and stared straight ahead. The forest was full of large nut pines, and the delicious smell of pine nuts wafted toward me. Suddenly, a face emerged from a nut pine five to six meters in front of me. I stared blankly as the contours of the face became clear. Its eyes were so deep, they looked like they were burning from the inside. It was a tiger.

The tiger stood very still with its eyes on me. I couldn't make a sound—or rather, I didn't dare. I couldn't even lift the camera hanging around my neck. The slightest move would prompt an attack. The tiger's head was enormous and its bone structure impressive. Its ruff was thick and its size was like nothing I'd ever seen. I knew instinctively that this was Khajain, the Great King, the spirit of the Sikhote-Alin Range. His eyes were at once aloof and piercing, and they were focused only on me. The look in his eyes wasn't one of someone who was caught, but someone who had come to see.

We stared at each other for a long time. He twitched his lips ever so slightly. It was an unspoken warning not to do anything foolish. That one twitch of the lip drained the last stores of energy I had left in my body. Khajain slowly walked out from behind the nut pine. I could now see his entire magnificent form from head to tail. As he slowly followed a diagonal trajectory from the nut pine to the mountain path, he did not once take his eyes off me. I still couldn't move a muscle, but I kept my eyes locked on his as though everything was fine. His gaze squarely on me, I felt psychologically paralyzed, and the tension sent shooting pains all over my body as though I were being stabbed with needles all over.

Khajain continued staring at me until he reached the mountain path, as if to communicate his determination to observe and understand everything about me. In that cool gaze, I felt a psychological distance that I could not willfully contract or expand. Once he reached the path, he turned away and walked on, his eyes forward. He did not once glance back at me before he disappeared into the forest. I felt ignored or disappointed. I returned to the small being I was the second Khajain took his eyes off me.

Khajain was traveling along a mountain path by Santago River. Over the thud-thud *of his own paws, he heard a human. A deer or boar would flee at that moment, but the Great King stopped in his tracks and searched the area. Great nut pines lined the mountain path. Khajain stepped behind one of them. It would just take a moment for the pesky humans to clear out, and he'd soon be able to continue on his way. That's what he had always done, and that's what he would do today.*

But the humans were not moving on, and Khajain was curious to find out why. He poked his head out from behind the tree to see two humans resting on a rock. One was lying down, and the other was sitting up. Khajain's eyes met the eyes of the human sitting up. He glared at him. He saw surprise and fear in the human's eyes, although the human attempted to hide it.

They didn't have any long sticks that smelled like metal. They were not hunters. Khajain did not feel hostility coming from them. Still, he twitched his lips subtly to warn the human. Then he got to his feet and slowly made his way out of the nut pine grove. With his eyes on the human, Khajain kept him bound at a safe distance and took note of his scent for future reference.

Once back on the mountain path, Khajain took his eyes off the human. He was not in the least afraid of him, though the human was still sitting behind him. The human was frozen stiff in his place, with fear in his eyes. Still, the perfectly focused gaze of a human being was always somewhat unnerving.

I woke Stefanovich. He stopped snoring and sat up as though he was never asleep to begin with. I told him about Khajain, and the veteran forest ranger's eyes became as wide as saucers. He ran over to the nut pine to look at the paw prints and expressed disappointment that I hadn't woken him. I told him I couldn't move or make a sound, that I thought Khajain would have attacked if I did. Stefanovich nodded. He'd once run into the legendary Great King, Kuchi Mapa, when he'd been walking alone in the woods.

What does it feel like to run into a tiger in the forest? One such scene appears in *The Great Van: The Life of a Manchurian Tiger*, a novel by Nicolas Baïkov, a Russian explorer. An old man comes across a tiger in the forest and has a fierce internal battle with himself. But on the outside, he seems perfectly fine as he continues on his way. It's extremely difficult to pull that off. I don't think I would have the guts to do it.

What I'd felt at that moment was a distance. A distance that needed to be maintained, neither shortened nor lengthened. I'd instinctively felt that the tiger was maintaining a certain distance and that I should, too. After that, a voice from deep within me had reverberated: *Don't make any sudden moves. Exercise self-control.* If I stuck to these rules, the tiger and I could safely go our separate ways. If I didn't, I would be attacked.

After the tiger had gone and I regained my mental facilities, there was one question on my mind. How did Khajain feel about our encounter? In Rudyard Kipling's *The Jungle Book*, Mowgli, the child raised by wolves, holds power in his gaze. When Mowgli stares down a pack of wolves, none of them can look back at him. Even the black panther, Mowgli's best friend, can't look him straight in the eye. This is because Mowgli's eyes are human.

The human eye can perceive an object only when the focus sits precisely at the center of the retina. In other words, to look at something, one must look straight at it. Most mammals, on the other hand, can see objects very clearly even from the corners of their eyes. This means that, except under extraordinary circumstances usually related to hunting, they don't have to look directly at something to focus on it. So when animals see other animals staring straight at them, they take this to mean they are being hunted. This, it goes without saying, makes them very nervous. The feeling is all the more intensified when it's a human eye staring them down.

Khajain must have felt something similar when we'd made eye contact. But he was able to meet my eyes, and his own gaze overpowered mine. He must have read the slight tremor of fear on my face, too. His gaze was cool but at the same time penetrating. I would expect no less from the Great King of Ussuri.

CHAPTER 24

Night of Beasts

WE TRAILED KHAJAIN'S tracks for two days and examined the path he had taken. His trail continued upstream along Santago River. We found more tiger traces as we traveled. At the source of the river, a young nut pine lay on the ground by a mountain path. Its leaves were stripped off and its trunk broken as though it had been knocked over and trampled. It was a very rare sight: it meant a tiger had lain on the pine and rubbed itself on it. Before long, we found a dropping stamp. The excrement mixed with yellow and reddish secretion sat on top of a little mound of dirt the tiger had made by digging the earth. A female tiger in heat had left it there. When a female tiger is in heat, she wanders around her territory leaving more marks than usual. She rubs her face and waist against trees, sprays urine, and rolls on the ground to spread the scent of a tiger ready to mate.

Tigers mate year-round, whenever they want to. We know this by the variety of different tiger cubs' paw sizes we find in the snow in the winter. They are born in different seasons. The mating season of each individual tiger depends on the ages and health of the tiger, the period of rest required after the previous litter is independent, and the amount of prey in the territory, but the seasons have no obvious influence.

Next to the female tiger's stamp was another stamp. The ground was dug so deep that the roots of nearby shrubs were exposed. This must have been Khajain. The Great King was pursuing a female tiger in heat. We followed the tracks to find traces of many tigers congregating. One was the Great King, the other was another male tiger with paws almost as big as Khajain's, and the third was the female tiger in heat.

The male tigers had left competitive claw marks and excrement piles. There was some blood in the midst of it. The male tigers had fought each other until they bled, while the female watched. Khajain had won and thereafter stayed close to the female. The other male had followed them at a distance and sometimes made his move on the female, prompting another attack from Khajain. The tiger had retreated far away and circled back. Yet another male had approached the female, only to be beaten out by Khajain as well.

The indigenous Udege call this period of tigers congregating in the woods every night the Night of Beasts. When a female is in heat, the males pick up her scent and flock to her. The males fight over her every night and sometimes bite each other to death. They spend one to two weeks courting the female and getting rid of competition. Once the Night of Beasts begins, the valley is full of the sounds of tigers roaring, treading on leaves, growling at each other, whining desperately, and breathing heavily out of exasperation. The sounds turn from irritated to sweet to hostile in seconds, but to human ears, it just sounds like the tigers have lost their minds.

The mating ritual intensifies as the female gets closer to her ovulation date and she starts to mate more frequently. A typical female mates sixty times in the two or three days of her ovulation. This is why people mistakenly believe that consuming male tiger genitalia will enhance their own sexual performance. In fact, *viagra* is Sanskrit for "tiger."

Female tigers in heat need to be stimulated for several days in order to ovulate, so even if a young male tiger manages to mate with her first, it's unlikely that he will succeed in impregnating her. When the strongest tiger in the area arrives, he chases the other males away and provides the female with the stimulation she needs to ovulate. On average, the stimulation and mating go on for two weeks. Another benefit of having cubs with the strongest male in the area is that the cubs will have a better chance at survival if they inherit genes from a strong father and grow up safely within his territory. So females tend to mate with the alpha tiger of the area, the Great King. Since Khajain was in the mix of contenders, he was probably the victor of the Night of Beasts. When I'd run into Khajain a few days earlier, he'd been on his way back after thoroughly enjoying his victory.

The female was probably either White Moon or White Snow. Female tigers become sexually mature at around thirty months. From that point on, they go into heat as long as they can secure a territory. White Moon and White Snow had inherited their territory from Bloody Mary and were almost three years old. It could be a female from outside the area, but it was more likely a female that frequented this territory—one of the two sisters. Like the leopards in Kedrovaya Pad, were the wild tigers of Lazovsky also resorting to incest?

THE FOREST WAS thick on both sides of Santago River. The water babbled as it broke into delicate white scales on the surface. The crimson pebbles in the shallows lay evenly, making patterns that glittered in the sunlight and reminding me of a turtle's shell.

Stefanovich and I turned a corner, and I saw a deer drinking on the other side of the river. In June, deer are often spotted near rivers adjacent to deciduous forests. They come down to the river to graze on the tender leaves in the morning or in the afternoon when the shadows grow long. The best habitats for tigers and the animals tigers feed on are oak groves and nut pine groves. Another ideal habitat is a dense forest such as this with a river nearby. Many animals live near rivers in forests, and tigers pursue them.

I took a closer look at the deer and saw that its behind was pointed up like a hopping kangaroo's. Its front legs were shorter than its hind legs. It was a musk deer. It looked up at me. A drop of water rolled down its long neck. Its incisors poked out sharply on both sides of its mouth. It was a male. The deer twitched its nose and pricked up its ears, then turned and ran down the pebble bank. Its fur was white from its curly tail all the way down to its thighs. The hunters of Ussuri refer to the musk deer's white rump as "the mirror" because it is easy to spot from far away and makes a good target. It's an endangered species in Ussuri because it has been overhunted for its scent gland. This one jumped a few times with its behind bouncing up and down and soon disappeared into the forest on the other side.

We traveled up Santago River to the coastal range via Shauka. Ten kilometers down the range, we came across a gorge. A small river flowed from the gorge into the sea. We broke our journey at the cliff. The long period of camping outside was starting to take its toll. We were running out of food and battery life. It was time to return to the base camp.

Globular shapes glistened in the river below us. Some of them peered out from under the water with only their hairless heads poking out, while others were diving. Some were on the coarse pebble beach enjoying the sun. Their plump gray bodies patterned with black dots were wet and shimmery in the afternoon light. *Nyerpa* in Ussuri, these were spotted seals.

Every female had had a pup born just that spring. The black dots were not yet vivid on these babies, and the gray of their coats was closer to white. These seals live up and down the Ussuri coast in places difficult for people to get to. They eat their fill of herring and Alaska pollock and then come to lie on the pebble beaches or climb on top of rocks to get some sun and rest. They are fast underwater but slow as slugs on land, which makes them anxious and vigilant on land compared to their curious, relaxed selves in the water.

A beast is never cross with the animal it pursues. It advances into the hunting range quietly and humbly, becoming stealthier with every move its prey makes. And so it was now. I looked down at the steep cliff and saw a tiger coming around the bend and ducking out of sight under the cliff in one fluid movement.

I held my breath and reached for my camera. The camera was halfway to my eye when I remembered it had run out of battery some time ago. It was always like this even when I tried to be careful. I picked up my binoculars instead.

The tiger slowly emerged from its hiding place. It was a small female tiger. She advanced carefully, minding every step, and hid behind a large fist-shaped rock in front of the cliff. She lay flat as soon as she got behind the rock. Her paws were firmly on the ground so she could spring forward at any moment. She craned her neck and spied on the seals. Each time she pricked her small ears, the white spots on the back of them looked like eyes staring at me.

In front of the fist-shaped rock was another shaped like a candlestick. The tiger jerked forward a few times, as if debating whether to stay hidden behind that rock. There was nothing but pebbles on the beach after the candlestick rock, which was less than thirty meters from the river estuary where the seals were resting. The hesitant tiger emerged from behind the fist-shaped rock. She got down as low as a four-legged animal possibly could and crept forward, then scurried the rest of the way to the candlestick rock, where she pressed herself close. There, she continued her patient wait for the right moment. I couldn't see the expression on her face, but the subtle trembling of her muscles showed that she was focused solely on the seals. Because of her intense focus to the front, she had no idea that I was looking down at her from the cliff behind.

She was tense like a loaded spring as she readied herself to attack. *Sorry, seal. I'm at the top of the food chain, and I'm getting hungry here. Law of nature, and all.*

She broke the tension and jumped. The seals didn't see the tiger until she had already made three leaps toward them. She was halfway there when they began to waddle into the water. The pups and a few midsize seals lagged behind. She caught up the moment the last seals hit the water. She didn't bother slowing down as she dove into the river. Drops of water flew into the air; one of the slower seals flapped around in the water. The seal's firm body thrashed wildly, splashing water everywhere. I saw streams of blood in the water. The tiger had the seal by its throat and was shaking it, trying to pull it onto land. The more the seal struggled, the harder the tiger throttled it, as if she was trying to be polite by killing it as quickly as possible.

The creature in the river finally calmed down. The tiger lifted her head and caught her breath. I saw fangs dripping red through her parted lips. Water dripped along her strong jawline and landed in the river over her submerged paw. A limp seal floated in front of her. Red flowed out of the seal, coloring the river.

It used to be a myth that tigers hunted seals. The Udege have a legend about the relationship between Amba and Nyerpa, but nothing had been confirmed until that moment. Scholars had found seal skulls quite far inland and speculated it was a tiger that had brought them there. When I told scholars what I'd seen, they were quite delighted. The Udege, on the other hand, were not. The descendants of people who have long lived on the East Sea know exactly when tigers hunt seals—when Temu, the great spirit of the ocean, regains his calm and finally stops sending rain, Amba, the god of the forest, hunts seals. The Udege ancestors know that the end of a long monsoon brings sunshine and lures seals onto dry ground, and the tigers know this, too.

The tiger took a few laps of water and dragged the seal behind the candlestick rock. It was a midsize seal that looked like it had been born the previous spring. The tiger plopped down and began eating. On the other side of the blood trail that led to the tiger, the more fortunate seals continued to bob up and down the surface of the water as they spied on the tiger.

The tiger was White Snow. She wasn't big, but she was confident. Her body was firm and her fur well groomed. She seemed to have completely settled down on the coastal range she'd inherited from Bloody Mary. She must have learned how to hunt seals from her mother. Living near the sea had probably honed her marine mammal-hunting skills. She had known where the seals would show up, had waited patiently, and had made an expert kill. It had been an amazing sight.

Feeling lucky to have witnessed something so rare, we quietly retreated from the cliff. As we hurried along, I got a feeling that the belle of the Night of Beasts may have been White Snow. A tiger's gestation period is about one hundred days. She would have her litter in August, at the end of the summer, or in September when autumn began.

CHAPTER 25

Chrysanthemums

THE CURRENTS OF Deer River ran gentle and long. The waters tumbling around the wide turns in the mountains ran slowly but eternally, pulling in the soul of the beholder. The countless ripples in the water danced and vibrated at their own frequency. The harmony was simple and unchanging.

"Do you want to try a radon bath?" Stefanovich asked as we walked upstream from Deer Valley.

"What's a radon bath?" I asked.

"There's a place for it around here. Let's go."

We arrived at a tributary of Deer River and found warm water flowing out of a steaming hole. The hole was yellowish and the ground around it was red. It turned out he was talking about a mineral hot spring.

Stefanovich told me that forest rangers sometimes stopped by this radon spring for a bath or to fill their canteens when they were in the area. There were deer tracks around the spring. They must have used them too, to replenish their minerals.

"Does it freeze over in the winter?" I asked.

"Only part of it," he said.

He told me the brook would freeze but that tepid water would continue to flow from a few spots, so animals could drink from the spring in the winter. I closely examined the surroundings and found some very old tiger tracks. We decided to put an extra bunker near the radon spring for the winter. The old forest ranger was surprised.

"You think you'll see tigers here?"

"When you're thirsty, do you want to wait for ice to melt, or gulp down warm water? I'll bet it's the same for animals."

"You have a point."

Every forest has a source of sodium or other minerals. Ungulates frequent these places to get their minerals. They would be at this hot spring even more often in the winter when everything was frozen. The track we found was old, but tigers had been here as well. They knew the springs existed. They must have also figured out that ungulates came here often. A tiger climbing the mountains in the winter must think of the radon spring when it's in the area. How wonderful to have a place where the vagabond on a long journey could warm himself with a sip from the hot spring and continue on his way.

Darkness spread. The moon rose shyly. The river was as bright under the moon as if it were dawn, and the fuzzy leaves of willows glittered like silver. When the air temperature dipped below that of the water, a mist rose along the river. The mist hung in a thin blanket over the water before dissipating. I gathered more wood and threw it in the fire. The fire burned red. The shadows of trees jerked each time a spark flared. Fish jumped occasionally and otters splashed. Silence fell again, and all I could hear was the sound of the currents sweeping the sandy riverbank.

I could see glimpses of Stefanovich's sideburns in the firelight. He looked old.

"Hey, *maladoy*."

"Yes, *stari*?"

Stefanovich insisted that people younger than him (even by a few years, even if they were over seventy) call him *stari* (old man), and he would call them *maladoy* (young man). It was a tradition among mountain folk. There was no distinction or prejudice in the forest besides chronology.

"Don't you want to go home, maladoy?" he asked.

"……"

"Don't you want to see your kids?"

"……"

"I keep thinking about my hometown these days. I want to see my grandkids, too."

"……"

I didn't say anything.

"I'm going to retire from my life in the mountains next year and return to my hometown."

Tears suddenly welled up in my eyes.

"You'll be a willow on a river one day," I said.

"And what about you?"

"I'll be a willow, too."

"Have you ever heard of Uza?"

"Yes."

Uza is a mythological figure whose name in Udege means "the person who cries" or "the person who gives birth to sadness."

The first Udege who lived on this earth didn't know what it meant to die. People believed that when they grew old, their spirits moved into trees. Men inhabited willows and women inhabited birches. The spirits lived in trees until they were born again, and thus they circulated indefinitely in this middleworld.

One day, Uza was walking along a lake in the forest. Something glistened on the bottom of the clear lake. It was a beautiful oyster.

"Uza, return to your house and dig under your brother's cot," said the oyster. Ripples appeared as it spoke. "You will find the entrance to *buni*, the underworld."

Uza ran home and dug under his brother's cot and found the underworld just as the oyster had predicted. In the underworld was a forest just like the one in the middleworld. A network of rivers flowed through the forest, and animals played. There was no sun, but the moon shone bright and cool. Uza returned from the underworld and covered the entrance with a birch mat. He was worried that when his brother, a famous hunter, found out about the underworld, he would kill all the animals there.

Evening came, and Uza's brother returned from a day of hunting. The brother sat down on the birch mat and fell into the underworld. Worried for his brother, Uza followed him. There was a large dog guarding the entrance to buni. It glared with terrifyingly large eyes and, when it saw Uza, barked deafeningly at him. An old man sitting next to the dog saw Uza and lit up.

"Uza!" said the old man. "I've been waiting for you a long time. You were the first to create this path to buni!"

A spirit that enters buni cannot return to the middleworld. The terrifying dog at the entrance prevented spirits from leaving. But the old man allowed Uza to enter the underworld and return to the middleworld. From then on, Uza became the guide that took the spirits of the dead from the middleworld to the underworld.

The circulation of spirits in the middleworld stopped after Uza opened the door to the underworld. There was death in the world after that point. People were born in great numbers, as numerous as there were trees in the forests. And with the passage of time and the guidance of Uza, they flowed into buni. There was sadness in the world.

"You asleep, maladoy?"

"No."

"How long have we been working together?"

"Seven, eight years?"

"How long have you been working in the mountains?"

"Almost twenty. What about you, stari?"

"Me? Fifty years, maybe..."

"Where would you rather be, up in the mountains, or at home?"

"I don't know."

"You should go home next year. Spend some time with your grandkids."

"You're right," he said. "You know... I don't much care for my plot in the cemetery. I don't know what it is about it, but I just don't like it."

"You'll become a willow on a river."

"You think so?"

A jealous cloud blotted out the moonlight. Before long, everything receded into darkness except for the campfire. The moon occasionally peeked out from between the clouds. The stronger the moonlight, the heavier the cold night dew. The campfire slowly died, and the moonlight flirted with the fading shadows. It was a cold, quiet night.

WHEN A LIFE ends in the forest, other lives come to pay their respects. Crows and eagles are the first to come. Then the raccoons and badgers. Finally, ants and carrion beetles come to take care of what remains. Once they've come and gone, the departed quietly melts back into the forest. But if no mourners come, the fallen life generates life from within to help him return to the forest.

Maybe the part of the forest where he fell was too remote for any mourner. Maybe he was left alone because he was a tiger. Either way, there wasn't a single crow in sight where he lay. Just swarms of maggots that hatched and fed on his body. They crawled in his nose and his mouth. Tens, perhaps hundreds of thousands of maggots formed mounds that broke down and consumed his flesh around the stomach and waist. Fated to live and die on corpses, the maggots climbed on top of each other. The ones underneath slid down like gruel.

White Sky was dead. He'd been strangled by a poaching wire.

We found him a month after his death on a ridge halfway up Crow Mountain between Deer Valley and Santago. I undid the wire tied to the bottom of the oak. He'd struggled so hard to free himself that the bark was peeled and the wire had dug into the trunk. His molars and fangs were still clenched when we found him. The wire would have tightened around his neck harder as he struggled. It was tiger poaching wire, famous for its strength—the same kind used to drag bulldozers stuck in the mud. The wire would have filled White Sky with despair until the moment he drew his very last breath.

We removed the wire from his neck and carried him down the slope. Rotten flesh fell off his body, each chunk crawling with maggots. His strong torso was reduced to one layer of skin. The bottom half of his body had already melted into the forest floor. A white film covered his once cool eyes, and maggots crawled over them. The mere sound of his footsteps had once made me feel alive, and the memory of his breathing when he'd attacked me in the bunker still gave me chills, but he no longer made a sound. The beautiful, wild White Sky who had stood between the azaleas on the rocky cliff with the waves crashing before him now lay dead and rotting before my eyes.

We gathered his remains. We stacked some dry branches and laid White Sky on top of them. His body caught the flame instantly when we lit the fire. He dripped as he burned, and the drops seeped into the forest floor. The smoke from his body hung above the Ussuri forest and spread across the sky. The flames looked weary. The fire grew ravenous and then slowly died down. We pulled the skull and larger bones from the ashes.

Deer River flowed ceaselessly, monotonous and unchanging, as we returned to the base. White Siberian chrysanthemums bloomed on the riverbank, set against the field of green leaves; lavender asters had come in as well. The fluttering blossoms seemed happy and innocent. Foxtails stretched their necks out from the rich bed of chrysanthemums. Tails fluttered at the tip of every thin, long reed. Beyond that, the pale clouds flowed on.

I thought of Dersu Uzala, an Udege hunter immortalized in the books of Vladimir Arsenyev, a famous Russian explorer from the early twentieth century. Vladimir had asked Dersu,

> "What's a star?"
> "That there is a star. You look at it."
> "What is the moon?"
> "A person with eyes sees the moon. That is the moon."
> "What is the sky?"
> "It's blue when it's bright. It's black when it's dark. It's gray when it's rainy."

PART V

Another Winter

CHAPTER 26

Hansel and Gretel

THE RIVER BEGAN to turn solid. The ripples on the bank made strange patterns as they turned into thin ice; these patterns were gradually obscured by the thick, opaque, silver mass of the ice. The freezing spread from the riverbank to the center of the river, and the current struggled to pioneer a new path between the sheets of ice closing in on both sides. When the thinning trickle was finally frozen in, the constant babble of the water grew still and Deer River turned into a white path. But the current always finds its way through the darkness of winter.

Then it snowed. Heaven and earth grew still, stiller than the night. The large trees on either side of the river reached up to grab each other's hands, and thick flakes of snow fell slowly, for a long time. The snow piled on branches and fell to the ground. If I listened carefully, I could hear the rustling of snowflakes brushing against each other. The sound of snow doesn't rush, but flows ceaselessly. In silence, snow buried the forest, the river, and the bunker.

I gazed out onto Deer River from the bunker entrance, my chin propped up. The more I looked, the more I was mesmerized by nature's white skin. I was overcome with longing for the freedom of the forest as I sat confined in my bunker. I wanted to roam through the snow, run and play until I collapsed. Was life like that snow?

Did it come down, pile up, freeze, melt, and then come down again? Another winter had begun, and I was deep into another stakeout.

In the forest to my left, the falling snow mingled gently with the snow blowing off the branches. An orange animal scuttled out of the flurries. It looked like a raccoon. But on second thought, its midsection was too long for it to be a raccoon. Its tail sloped at a forty-five-degree angle and then curled up at the end; its ears were small, and its eyes were determined but oddly sleepy at the same time. Its face was young and full of mischief. It was a tiger cub. It had appeared in this deserted forest in the first snow of the year in Deer Valley. About ten meters behind it, another orange ball of fur trod carefully out of the bushes. There were two of them! Against the large, white trees shedding piles of snow, their young bodies were orange and adorable. It felt like a dream to be seeing them before me.

The first cub courageously came out of the forest and stepped onto Deer River, which had turned into a snowy tunnel. Snow-covered branches formed the roof, and the velvety snow on the surface of the ice was the cave's floor. The cub stood in the middle of the empty snow cave and gazed around curiously.

Siberian tigers generally give birth to two or three cubs. Very rarely do you find a litter of all males or all females. There's less hierarchy among siblings compared to other animals. Male tiger cubs tend to be more courageous and outgoing, whereas the female is more cautious. Female cubs prefer to let their male siblings explore unfamiliar territories first before joining them. I thought that the one that stepped onto the frozen river must be male. His sister lay flat on her stomach behind a bush that looked like a bouquet of snow flowers and watched her brother. They looked like fairytale siblings lost in the woods. I named them Hansel and Gretel.

Hansel took half a dozen brave steps, then stopped. Surrounded by snow, the river felt as intimate as a cave, but it was still in the middle of an exposed area. He may have been a cub, but he had a tiger's prudence. He took a few more sprightly steps, then buried his face in the snow and drank something. He had found an unfrozen radon spring. The entire river was frozen and buried under snow, but there were a few small holes thanks to the hot springs. The cubs were here to drink the water, but how had they got here without their mother? And how did they know there was an unfrozen hole here?

Wild tigers leave the den with their mother when they're two months old. At this point, their bones are strong enough for them to follow their mother around. Once they head out into the world, their nomadic life as wild tigers begins. Judging from their size and behavior, I estimated that these cubs were about three to four months old. They had probably been living as nomads for a little over a month now.

Hansel lifted his head from the water hole. He licked his lips with his long tongue, contented with the sweet drink of water, and looked around wide-eyed at the snowy world. He appeared proud and confident, like a child in a schoolyard who had won all the marbles.

Bolstered by watching Hansel safely drink water, Gretel carefully came out of the bushes, made her way to the old tree that had fallen across the river, and sat. She craned her neck and looked at her brother. Hansel came near the fallen tree, and she shrank, her ears pricking. Hansel nudged her with his snout as if to say it was her turn now, and she quickly got up to rub her cheek against his. She tapped him on the face with her tail and gingerly advanced to the water hole.

Suddenly, a tree branch that could no longer bear the weight of the snow bent and let its load fall to the ground. Gretel froze. With the water hole now only one to two meters away, she stayed still. Only when Hansel came bounding up playfully did she take her snout to the water hole. Unlike her brother, she looked up from the hole once in a while to keep an eye on things as she drank.

Gretel was still drinking water when Hansel went behind her and sniffed her bottom. The startled Gretel plopped down in the snow. Hansel climbed on top of her and tried to wrestle, but she would have none of it. She lay flat on the snow and looked around in bewilderment. Whether it was because she was in the middle of an exposed river, or because she had never seen snow in her life, or because their mother wasn't with them, Gretel was sensitive and cautious. When she refused to play, Hansel gave up and played in the snow by himself. He was deeply immersed in the fun of frolicking in snow, licking and digging, running and rolling.

A crow lit on a branch and cawed. It hopped up and down the length of the branch and watched the cubs. Snowflakes fluttered each time the branch shook. Hansel gazed up at the black bird that stared at him, intermittently making that awful sound. His eyes seemed to ask, *What kind of creature is that?* He knew he couldn't catch it, but he jumped up the trunk of the tree the crow was sitting in. The trunk was straight, and the cub made it only about two meters before he slid down again. The crow looked down at the cub jumping up and down and making a fool of himself. It rotated its head left and right and cawed.

Hansel reminded me of a cub I had seen a long time ago by the river. A flock of butterflies had been sipping water in a marshy area next to the river after fluttering from flower to flower. A tiger cub had come out of the forest, snuck up on the butterflies, and pounced on them. The butterflies had flared up like a large flame. The cub had jumped up and down along the river trying to catch them, and then disappeared into the forest. Like that cub, here was Hansel on a snowy river hopping around trying to catch a crow. The sight of a tiger cub playing without its mother was like looking at a happy baby laughing to itself in a crib.

I waited quietly, thinking the mother would appear any minute, but she did not. She must have gone to hunt. When female tigers with young cubs go hunting, they hide their cubs in a safe place. They stalk ungulates alone, and when the hunt is successful, they drag the prey to a safe place and return to fetch their cubs.

But cubs don't stay where their mother leaves them. They wander through the nearby forest looking for adventure. When they get sick of playing with each other, they show interest in everything in the forest. They wander some distance from their hiding place as they taste and scratch everything in sight. But they can never get very far. Even if they do, their mother can use her sharp senses of smell and hearing to hunt them down quickly. She then takes the cubs to the spot where she's hidden the prey, and the family has a feast. This nomadic lifestyle continues until the cubs are independent. Its days as a cub are the most carefree time in the life of a tiger, which spends its adult life alone with no one to depend on.

I had no way of telling if Hansel and Gretel were conceived during the Night of Beasts I'd heard months before. But I assumed their mother was White Moon. Deer Valley was in White Moon's territory, the inland area she'd inherited from Bloody Mary.

Somewhere, a black woodpecker pecked at a dead tree with its solid beak. Amplified by the hollow space inside the dry wood, the sound resembled the wooden hand drum monks play at Buddhist temples. Hansel got up. The sound must have unnerved him. He traced his steps back into the forest. Gretel quickly took a few more sips of water and hurried on to catch up with her brother. Unlike her careful movement when she'd come down to the river, now she scampered. The cubs blinked between the trunks of large trees and disappeared into the snow.

All that remained was the hollow sound of the bird pecking wood.

CHAPTER 27

The Future of Ussuri Forest

With just a few days remaining until the new year, the long-awaited supplies finally arrived. Supply days came around once every two or three months and punctuated the long stakeout period like teatimes. The delivery person would bring three hundred rice balls, salt, batteries, and—my favorite—things to read. The excrements box would be sent out, and I would receive a new supply of water.

Stefanovich looked tired after hauling the heavy supplies through the mountains. We sat in the crowded bunker together, made tea, ate rice balls, and caught up.

Stefanovich said he'd seen tiger tracks a few days earlier. Two cubs had gone over the coastal range, but there were no signs of their mother. I was surprised to hear this. If they were found in the coastal range, did this mean White Snow had had cubs, too? Stefanovich was also surprised to hear that I'd seen two cubs in Deer Valley that might be White Moon's.

White Moon's main territory included the Dragon Spine, located inland, and ranged from Black Mountain all the way down to the south of Crow Mountain. White Snow, on the other hand, had settled down on the coastal range after White Sky's death.

The tracks didn't reveal any territory overlap, but Bloody Mary had traveled from the Dragon Spine to the coastal range when she was raising her cubs, so there was no guarantee that White Moon would not occasionally take her cubs to the coastal range as well. If that was the case, then Stefanovich and I were talking about the same cubs. If not, White Snow also had cubs.

I wanted to go look at the tracks myself. Forest rangers like Stefanovich have an excellent grasp on the overall ecosystem of forests, but may not know everything about the minute details of one specific species. Tiger cub tracks are so similar to Amur leopard or lynx tracks that, upon first glance, they look the same. If Stefanovich had indeed found White Snow's cubs, we had to adjust our stakeout plans. I put my Deer Valley stakeout on pause and followed Stefanovich to the coastal range.

He took me to a spot forty kilometers away from Deer Valley. There were tracks on the path to Triparashonka Beach. The tracks belonged to two cats, one with a paw pad width of 5.9 centimeters, the other 6.4 centimeters. Overall, the paw prints were the shape of apricot blossoms, and the absence of claw marks pointed to a big cat. I knew they weren't lynx, because the stride was long relative to the size of the paw print. Lynx have higher centers of gravity and shorter bodies. These prints were either from tiger cubs or leopards. If they were tigers, they were young cubs of four to five months, and if they were leopards, they were a fully grown male and female. An adult female leopard has a paw pad width of 6 centimeters, and a male 7 centimeters. Some leopards can have paw pads up to 7.5 centimeters wide, but those are very rare.

The tracks headed up the coastal range and met another set at the ridge—9.5 centimeters, an adult female's paw pad. The mother had met up with her cubs here. The paw sizes suggested these cubs had been born a little earlier than White Moon's. They probably belonged to White Snow, who lived on the coastal range. White

Snow, the very same who used to trail White Sky, had had cubs of her own. Time passes and once-helpless children become adults.

White Snow had led her cubs over a ridge and down into a valley of oak trees dotted with the odd nut pine and rock. The valley was sequestered and cozy. There was a small spring where animals would have come to drink before it froze over. The spring had frozen over so many times that it had formed an enormous ice patch. It extended to the cliff's edge, where it turned into fat icicles over one meter long.

The trunks of the oaks around the ice patch were bare up to chest height. This must have been related to last fall's poor acorn production. The acorn yield had been 20 percent less than the previous year. Many ungulates had left the coastal range for other regions with more food, and the deer and wild boars that remained struggled to get through the cold winter by peeling and eating tree bark.

Farther down the ice patch, we found a young fawn shaking in the snow. It had been born in the early summer. The mother was nowhere to be found. The lone fawn looked up at us with its sad eyes. It was so cold and hungry that it couldn't muster the strength to get up when I walked up to it. If we were to leave it here, it would die of starvation and be buried by snow. It would become crow and eagle food by spring. I felt sad for the young life that would never get to bloom. I pulled out a piece of black bread from my backpack and put it in front of the fawn. It took a few bites. I left a second piece of bread and continued to follow White Snow's tracks.

At the bottom of the valley, we found what remained of a wild boar carcass. There were tiger tracks everywhere. White Snow must have hunted a boar somewhere and brought it here. And then she brought the cubs to eat it. Everything had been devoured except for the head and feet. The boar's ankles and hooves were hatched with fine tooth marks. The cubs had gnawed at them even though they were nothing but skin and bone.

The mother tiger wouldn't touch ungulate hooves because she knew from experience that there was nothing to eat there. But at four to five months old, the cubs were in the process of growing teeth and claws and liked to scratch and chew anything in sight. From this period on, their adult teeth would start to come in, and their baby teeth would be completely replaced with adult ones by sixteen months. They would start hunting large ungulates once they had their new teeth. They would hide their sharp teeth and claws until the right moment, then plunge them deep in the bodies of their prey.

Tigers have thirty teeth altogether, but the four fangs on the top and bottom of the jaw are especially important—they can hardly live without them. Hurting their fangs—while growing in their adult teeth, for example—can lead to serious problems. Tigers with injured fangs may start to attack livestock because they can't hunt large wild animals, and eventually they may hurt people, too. So a cub's secondary dentition is very important to its future.

The claw and tooth marks on the hooves showed that White Snow's cubs were growing healthy and strong, but I wondered if the excessive chewing also meant they weren't getting enough food. I worried about the poor acorn yield that year.

After feasting on the boar, White Snow had headed back to the top of the coastal range. Enormous rocks and coniferous trees stood along the path at the top and framed the East Sea vista. Strong winds blew across the undulating blue waves. The wind carried the refreshing scent of the ocean over the span of the temperate forest right to the foot of the Dragon Spine, which I could almost see. White Snow and her cubs had gone south along the ridge.

We followed White Snow's tracks for three or four kilometers and found another set of prints, which had come up from the mountainside to join them. The enormous paw pads (12.9 centimeters!) showed it was Khajain, the Great King. The very Khajain who had held a staring match with me from behind a large nut pine had traveled the vast and majestic Sikhote-Alin, stealthily as the wind, and returned after six months and over two hundred kilometers.

Did he happen to run into them, or had he come to say hello to his daughter? Male tigers are known to visit their cubs and their cubs' mother, if only once every few months, so it was likely he had come by for a visit. But White Snow was an adult, and tigers seldom visit their children once they're fully independent. Did that mean he had returned because White Snow's cubs were also his own?

Khajain had walked along the mountain path with the cubs. He had led and the cubs had followed. The small paw prints inside the large prints were adorable. If they weren't related, they wouldn't have been so friendly with one another. If her cubs had been fathered by another tiger, it wouldn't have mattered that White Snow was Khajain's daughter. Tragedy would have ensued. I thought, then, that the lucky couple on the Night of Beasts the previous summer had in fact been Khajain and White Snow—the father mating with his daughter. It may seem normal in nature, but it's actually sad.

Logging and development are isolating tiger habitats, and the number of wild tigers is decreasing in these fractured places. Tigers have no choice but to mate with their blood relatives, but they can't produce cubs with good genes if inbreeding continues. Three or four generations of inbreeding can lead to stillbirths; the cubs that are born might survive infancy, but they could fail to reach full mental capacity or sexual maturity.

If the inbreeding stopped after one generation, it would be okay. But when I thought of the future of the 350 Siberian tigers remaining across the Korean Peninsula, Manchuria, and Ussuri, I couldn't help but wonder if what I was seeing now was the future of the Ussuri forest. What if White Snow's cubs met tragic ends? My worry grew as I followed the tracks of White Snow's family along the mountain ridge.

CHAPTER 28

The Great King Comes Home

ANOTHER YEAR WAS coming to a close. Seven days after I moved to the bunker in Diplyak to observe Khajain and White Snow's family, a herd of deer came over the hill. Half a dozen of them walked down to eat the seaweed on the beach. Their hooves submerged in the slushy sea water, they fed on the marine plants on the rocks, too. Each time the waves hit a rock and water rained down like beads of ice, the deer shrank with fear.

The deer had been foraging for food on the beach for several days. There were no acorns on the oaks in this part of the coastal range, either. Even deer that spend their lives watching vigilantly for predators cannot help but take the risk of being seen on the beach to feed on marine plants. This is what happens during a bad acorn year and a harsh winter. In the evening, they retired to the cozy oak grove in the coastal basin to spend the night.

An Ussuri sika deer in its prime has antlers with five branches. The leader of the herd currently staying in the coastal range was a male with antlers like that. "Five-Prong" had led the herd out to the beach once again this morning. After he'd had his fill, Five-Prong headed back to the oak grove, his impressive back muscles contracting and expanding. The others followed.

At that moment, a tiger flew down from the hill surrounding the coastal basin. The deer fled in all directions. No one had time to give a warning call. Five-Prong, who was at the head of the pack, was the closest target. The tiger whizzed through the bare winter trees and bolted at Five-Prong. The trees quivered in the tiger's wake. The deer contracted and expanded every muscle in his body to get away, but all he could manage to do was get his long legs stuck in deep snowbanks. The tiger dashed across the snowy field by folding and releasing his back like a spring. With each leap, a small gust of wind formed a long groove in the snow.

The orange tiger was an unbound spirit that relied on no one as he raced through life. He seemed heroic. The scene made me think of the cavalry of Genghis Khan racing through the Gobi Desert with a plume of sandy dust trailing behind. This creature was a great warrior, too.

The brief duration of the chase felt like an eternity. The tiger closed in on the deer and took the final leap. They rolled in the snow together. With both of them buried, all I could see was snow flying up. The deer's long scream echoed through the cold forest until it abruptly stopped. The tiger must have snapped the deer's neck, killing it instantly. The forest was still again. Soon, a pile of snow rose. Some of the snow fell off to reveal a tiger still largely covered in it.

When wolves hunt, they economize by spreading out their limited energy across a long period of time. Endurance is their game. They pursue their prey until it gets exhausted. To make best use of their endurance, they work as a team under the command of their pack leader. Tigers, on the other hand, make the best use of their limited energy by expending it all at once on speed. This strategy requires furtive stalking, hiding, sprinting to catch up with the prey quickly, and using power to bring down a large animal like a deer or boar. Tigers developed this hunting strategy because they are a species that does not hunt or live in packs.

The oak grove lay under fifty centimeters of snow. The deeper the snow, the greater the tiger's advantage in hunting deer. Deer can run faster when there's little to no snow on the ground, but tigers have the speed advantage when there's a lot of snow because their powerful kicks shovel it out of the way. But speed alone doesn't get the tiger very far. For a successful hunt, the tiger must also get as close as possible to the prey without being seen. This can be tricky in the bare winter forest, so tigers wait until after sundown to hunt. Nature is always fair.

But this tiger hunted the deer during the day. It was lying in wait for the deer to return from the beach. It also took advantage of the wind direction. It blew toward the sea at night and toward the land during the day, so the tiger hid his scent from the deer by hiding upwind. This was an experienced tiger.

The tiger shook off the rest of the snow. It was too far away for me to see well, but his silhouette was large. Judging by the fact that he'd snapped the neck of a deer in its prime, I was sure this was a male tiger. Now the tiger dragged the deer by the neck and placed it under the young pines where White Snow and White Sky had squabbled over a deer the previous winter. The tiger slowly looked around the pine grove, sniffing the air and taking everything in. There were vivid markings on his body. Prominent shoulders. He lifted his head. He had a full ruff and a very large head. He was the Great King.

Khajain lifted his head up high and made a threatening and awful face. He wrinkled his nose and pulled his lips all the way back to reveal his gums. I saw his large fangs. He then breathed through his nose and took in the scent of the air. He breathed in the forest to remind himself that he was home. White puffs of breath came out of his nose. This was a behavior I called "homesicking," more technically referred to as the Flehmen response; it's what tigers do when they're home or at a place where they feel as safe as if they were home. It's similar to sailors breathing in the smell of land after a long, difficult voyage.

Tigers pick up scents through their gums as well as their noses. Just as the whiskers of a tiger help it balance and maintain a good sense of direction, its gums are an important sensory organ that pick up scent and taste. Tigers primarily use their noses to smell things when they're engaged in everyday activities such as hunting or avoiding unwelcome guests. But when they feel they're at an especially cozy place or when a male tiger picks up the scent of a female, they take in the scent and the taste by using their noses and their gums.

Zoo tigers sometimes make this face, too, when they smell a female or when they are full and happy in the breeding center where they were born and raised. But people think that it's frightening when a tiger makes this face. It's also a little strange when a zoo tiger is homesicking like a wild tiger. It almost looks like he is lamenting his confinement and longing for his home out in the wild.

Khajain scrunched his face several times, but his heart was content. He was pleased that he had hunted a large stag and was content as he recalled his happy memories of this young pine grove. Two years earlier, Khajain had met Bloody Mary on this coast and shared a meal with his cubs, White Moon, White Snow, and White Sky.

Khajain had grown up on the southern coastal range and the Dragon Spine until he'd usurped the old Great King, Tail, and taken over all of Lazovsky. Where I was now may have been Khajain's hometown. Long ago, he could have played with his siblings here when the wild cherry trees along the brook were in full bloom and drank the cool water where petals floated. Perhaps Khajain returned to the coastal range every six months or so because he longed for his home when he was on the road.

Khajain stopped homesicking and lay down, looking comfortable. He ripped out the deer's sphincter and devoured its steaming insides. He then used his front teeth to pull out the fur on the deer's rump. Once most of the fur was gone, he licked the bare skin. Using the rough needle-like papillae on his tongue, he pulled out the rest of the fur and started eating. Each time he chewed, a chunk of flesh was ripped off and I heard the powerful sound of his teeth crushing deer bones. With his head lowered, I had a clear view of the back of his head and his back. The black stripes spelled 王 (king) and 大 (great). He was truly a Great King in his prime.

Khajain seemed to be full. He licked his lips and groomed himself all over. Animals in the cat family are obsessed with cleanliness. They don't lie or even step in filthy places, and if they get something on their body, they use their tongue to lick it off. Their saliva has antibacterial properties as well as cleaning properties. Wild tigers groom themselves both to protect their bodies from germs and minimize their body odor to hide themselves from their prey.

Many zoo tigers, on the other hand, do not maintain the same level of cleanliness. The biggest difference between zoo tigers and wild tigers is their fur. Wild tigers have a clean, glossy coat. Both white and orange parts are vivid. They groom themselves every chance they get. But zoo tigers look shabby even though they're well fed. They sit anywhere, and their fur isn't groomed. It's dirty and oily. They don't feel the need to minimize their body scent by keeping themselves clean.

Khajain licked the back of his paw and rubbed his face and forehead with it. He looked as cute as a housecat. Then suddenly, he rose to his feet and listened to the sounds around him without moving a muscle. I heard waves, wind, reeds brushing against each other, and a woodpecker pecking a tree in search of dinner. Khajain relaxed again. He shook his body and then each foot, one by one. Then he slowly headed up the mountain.

THE MOON SAT low on the horizon. The Eurasian eagle owls were back this year. I could hear them calling and responding. But their hooting was silenced by the roar of a tiger echoing through the coastal basin.

I quietly switched on the night camera. Khajain stood next to the deer he had killed earlier in the day. He turned to the forest and roared again. Tigers never roar except during mating season and when they're calling their family. They let out a short, thick roar or a low growl to threaten or warn other animals, but those aren't really roars. The roar of an Ussuri tiger echoes through the mountains like a harrowing cry, but few people have ever heard it—neither villagers who live in tiger-inhabited areas nor tiger researchers. Tigers roar only in the middle of nowhere, where there is little chance of exposing themselves to danger.

In the dark forest, another tiger roared back. They exchanged several calls and responses, and then two cubs came bounding out of the woods. A female tiger followed them at a leisurely pace. It was White Snow.

The cubs dove into the deer as soon as they arrived. White Snow sat gently and took a few bites too. The cubs ate the rump that Khajain had been eating, and White Snow took the chest. Khajain sat with his front paws together and watched them. I heard the ribs snapping. Once White Snow had had enough, she switched spots with Khajain. Remarkably, the four of them never ate at the same time. There was always one sitting apart to keep watch.

After their meal, the family members sat and cleaned themselves. They wiped the grease and bits of food off their lips and paws, tidied their ruffs, and combed the fur on their shoulders. White Snow used her big tongue to clean the cubs' faces and backs where they couldn't reach.

One cub that appeared to be male jumped onto the Great King's chest as he lay on his back and gnawed on his great front paw as if it were a rabbit. Then he grabbed his father's paw with his front paws and kicked at it with his hind legs. The other cub playing with Khajain's tail looked female. Khajain waved his tail like a slithering snake and teased the cub. The cub tried to catch the tail, pouncing on it, standing up on its hind legs like a bear, and then falling over. Like a father who didn't want to discourage his child, Khajain lowered his tail and let the cub grab it. The other cub, now tired of playing rabbit with Khajain's paw, played with White Snow's tail and then jumped up on Khajain's back. Khajain bared his great fangs at the cub and pretended to bite it, and then played horse with the cub.

I named the female cub Mapa and the male cub Kuchi Mapa. Mapa is the totem of an Ussuri tribe that worships bears, and Kuchi Mapa is the totem of one that worships tigers. Kuchi Mapa was also the name of a legendary Great King of Ussuri. As I watched young Kuchi Mapa playing horse on his father's back, I hoped he, too, would grow up to be a Great King someday.

Khajain patiently played with his cubs for a long time. The moon started to sink below the clouds. The Eurasian eagle owl was quiet—it must have fallen asleep. The basin was silent. Khajain got up and stretched his front and back legs. Then he raised his head high, wrinkled his nose, rolled back his lips, and took in a deep breath with his head turning from side to side. He collected the memories of this forest. He did this for some time, and then walked up to White Snow and nuzzled at her neck, making a sound similar to a horse's whinny. Then he raised his tail, sprayed the base of a young pine with his urine, and continued on to a small brook nearby.

On both sides of the brook, big wild cherry trees grew, leaning slightly toward the water. Khajain quietly walked between the rows of trees. The cubs started after him, but stopped when they saw that their mother wasn't coming along. Khajain walked away without looking back, and White Snow watched him go. He became a faint silhouette and finally disappeared into the dark. All that remained were the gray shadows of the cherry trees.

CHAPTER 29

Raising Hansel and Gretel

BROWN PATCHES STARTED to appear in the snowy Deer Valley. The snow had melted in a few sunny spots to reveal leaves. In the middle of the white ice sheet growing more and more holes, Deer River crawled on like a centipede. About one hundred meters from the riverbank on the opposite side of the river from the bunker, large trees and rocks were perched on a slope. Next to them was an animal path that you couldn't see unless you were carefully searching for it. It was the path leading to the river from the forest. An old tree had fallen across the path, blocking it. The tree was now covered with black moss that would have been green in the summer. Clusters of mushrooms were growing on it. They looked like either black hoof mushrooms or reishi mushrooms. One of the clusters must have been very old because it was extremely big and round. Through my camera, I stared at the mushroom and, to my surprise, saw that its flaps twitched occasionally.

I zoomed in on the mushroom. The hazy camera monitor awoke and displayed the round mushroom. The mushroom looked like it had ears and was pricking them. It even had glowing eyes. It wasn't actually a mushroom: it was a tiger's head. It was peeking out at Deer River from behind the log. It was a female tiger, probably White Moon, since I was in her territory.

It was the perfect disguise. The brownish hill where the snow had melted was the approximate color of a tiger. The trees and the rocks were covered in dried moss that had the texture of fur, and there were brown leaves where the snow had melted. White Moon had hidden herself in surroundings that resembled her own fur to stalk her prey. She looked like she was ready to jump out at any second. But Deer River was silent. Sunlight alone glittered on the silvery ice path—not a deer in sight, not even a raccoon. The flat, wide face of an owl looked down at White Moon from a branch overhead.

From the forest on the other side of the hill, the sounds of twigs snapping brought a tiger cub running out of the forest. Another followed and pounced on the first one. They wrestled each other for a while, and just as the one on the bottom was about to pin the one on top, the latter hopped up a nearby tree. The trunk of the tree was at a forty-five-degree angle. Perfect for tiger cubs to climb.

The other cub hurried up the tree in pursuit. They had a confrontation in the middle of the trunk, slapping each other with their front paws and trying to bite the slapping paws. One of them hit the other squarely in the face and then took off. The one who was hit grumpily watched the one running away, and peeled off a piece of mossy tree bark to chew on. Hansel and Gretel. White Moon's two cubs had grown by leaps and bounds in the two months since I'd seen them.

White Moon surveyed the area, trying to decide if she should get up or not. Ignoring their mother's careful stalking, the cubs ran around the forest being their rambunctious selves. White Moon turned to look at her silly cubs. Her tense eyes and folded-back ears dissolved into a look of resignation. It didn't matter how good she was—there was no chance she would catch even the dimmest prey as long as these two were around. White Moon got up. The tawny owl flew off.

White Moon carefully felt her way down to Deer River. The two cubs ran downhill past her and landed on the water. Showing none of the timidity of when they'd been here on their own, they dashed to the water hole and gulped down the water. White Moon took a sip, too. The wide-open terrain must have made her nervous, for she moved to the hillside as soon as she was done drinking. It was only a few steps away, but she seemed to prefer the black and brown hill to the white, frozen river.

Now that their mother was with them, Gretel was as gutsy as her brother. She fought back to the end when Hansel pinned her down. Hansel sneaked away and scaled the steep slope on the other side of the river, and Gretel chased him up the hill. The slope faced north, so the snow there hadn't melted yet. They rolled down the hill and wrestled each other in the snow. Once they reached the bottom of the hill, they climbed up again. They rolled down and climbed back up several times. Brother and sister enjoyed themselves like children on sleighs.

White Moon's concerned gaze never moved away from her cubs. She looked relaxed, but I saw flashes of anxiety in her eyes. If she heard so much as the flutter of a bird wing, she whipped her head around to stare at the source of the sound. Ussuri tigers' vigilance goes beyond anything we humans can imagine. Female tigers with cubs are especially dangerous. Their maternal instincts are in overdrive.

It seemed like only yesterday when White Moon, White Snow, and White Sky were trailing behind Bloody Mary. It felt like no time had passed since Bloody Mary's death, but White Moon was now a mother. Her frequent appearances with the cubs in Deer Valley suggested that she had given birth to them on Crow Mountain.

When a female tiger is ready to have cubs, she starts searching the forest for a den in which to give birth and raise the cubs. A first-time mother may prove lazy or incompetent at it and give birth under a clump of dirt clinging to the roots of an uprooted tree instead, or worse, in the bushes. But experienced tigers are very picky about their dens. The den must provide a good environment to raise cubs, and it must be the sort of place where the cubs won't be harmed by humans or other beasts when she's out hunting. Tigers like dry, south-facing caves shielded from the wind and rain, with a decent hunting ground like Deer Valley nearby—it's even better if the spot is hard for unwanted guests to access.

Once the tiger has chosen her den, she's careful not to let humans or other beasts find out where it is. She avoids taking direct paths there, throwing potential enemies off the scent by taking long, roundabout routes, and tries to leave as few telltale tracks as possible. She does not leave any territory markers like scratch marks or excrement near the cave, because when a tiger is pregnant, everything in the world is a potential enemy—even fellow tigers.

When a tiger gives birth, she stays in the den for one week looking after the newborn cubs. After a week, she goes out to drink water and hunt. When the cubs grow teeth, she weans them off her own milk and starts feeding them small animals like pheasants, rabbits, and badgers. Once the cubs become accustomed to meat, she shows them how to take apart a dead animal and has them eat the meat off the bones on their own. Once they learn to do this, she brings them live animals and sees what they do with them. If the cubs can't kill the animals, she shows them how.

Once the cubs are big enough to leave the den, they follow their mother wherever she goes. The cubs learn about the forest as they travel with her—paths, shortcuts, dangerous places, safe places, places for quick rests, places for long rests, places where the hot springs do not freeze over during the winter, places where prey appears regardless of season, and so on. Once this basic education is done, they learn more advanced things like what to do when a human appears, how to avoid manmade structures, how to cross a road, how to tell the difference between wild animals and livestock, and how to clean wounds.

The cubs also learn how to hunt. The mother tiger leaves her cubs in a secluded place and hunts alone at first, but as the cubs grow older, she brings them with her. She has the cubs hunt small animals on their own. Cubs often miss their prey because they don't have the patience to wait for it to get close enough. Through trial and error, they learn to wait and to move soundlessly. Then they learn about different animals—how to hunt birds, how to fish, where and when seals appear, how to deal with poisonous snakes, that hedgehog needles hurt, that raccoons and badgers play dead when they're cornered, that raccoon meat smells bad and isn't very tasty, that badgers are tasty but have a temper and a long, sickle-like nail that can really cut you, and so on. Cubs learn the basic lessons by being around their mother, but the more advanced skills require demonstration and training.

Cubs get their full set of adult teeth by sixteen months. At this point, their fragile claws become stronger as well. They begin to hunt large ungulates like deer and boars. They start out with younger or sicklier animals, but the average tiger can hunt a large ungulate on its own when it is eighteen months old. Their success rate, however, is not as high as their mother's. The real hunting training doesn't begin until cubs start hunting large animals. Cubs can get seriously injured going after large boars or bears without fully understanding what they're getting into. This isn't the sort of thing their mother can teach them. They must face these challenges head-on and learn from them. After about a year of semi-independence, during which they follow their mother but experience life on their own, the cubs are finally ready to live alone.

Tigers are usually very sweet to their cubs, but strict when it comes to their education. For example, when White Snow's family met Khajain at Diplyak, White Snow kept watch as the cubs ate their fill of the deer Khajain had caught for them before he left. When it was White Snow's turn to eat and the cubs' turn to keep watch, the cubs monkeyed around instead of keeping an eye out. Nevertheless, the mother and the cubs took turns "keeping watch." When it was White Snow's turn to eat again, the cubs played around by the deer instead of heading up the slope to serve as lookouts. White Snow thought that they wanted to eat more, so she returned to her post, but the cubs kept fooling around instead of eating. White Snow let them get away with it twice, but the third time was unforgivable. She growled menacingly at the cubs as though she was ready to rip them to pieces. The frightened cubs clambered up the hill to keep watch. She had to teach them that when they eat, they take turns. And that when their mother eats, they have to be on the lookout, too.

After running all over the snowy hillside, Gretel returned to her mother. Exhausted from all that sledding, she threw herself into her mother's embrace. White Moon licked the snowflakes off her daughter. Hansel stood on top of the hill and looked down at his mother and sister from afar. He seemed disappointed that no one would play with him. He dragged himself back, resigned, and then suddenly pounced on Gretel again. Gretel wasn't going to take it lying down. They raised their front paws like wild horses and slapped each other around. One of them fell into their mother. White Moon got up and walked away from her raucous children, back into the forest. The brother and sister were too busy fighting to notice their mother leave.

White Moon called them, a strange combination of a yawn and a roar, from the forest. Gretel stopped hitting her brother and ran over to her mother. Hansel, once again without a sparring partner, looked blankly at his sister as he licked his nose with his long tongue. Then he dashed at Gretel, knocked her down, and ran away. Gretel followed him. Enjoying to the fullest the fact that they were born as siblings, they played an endless game of tag. Keeping them safe within her peripheral vision, White Moon slowly disappeared into the forest. She had her work cut out for her for the next three years. It would be no small task staying watchful until her cubs were all grown up.

CHAPTER 30

Requiem for the Fading

"HELLO, HELLO. THIS is Petrova! There's a tiger circling the lodge. Chara is gone! I repeat. Tiger at the Petrova Lodge. Chara missing."

The two-way radio crackled urgently in the early dawn. The second I heard the message, I thought of White Snow. The image of White Snow from the previous year when she had stalked Chara, the lodge dog, with White Sky overlapped with an image I had of Tail, the Great King. Before he died, the elderly Tail had resorted to attacking livestock.

In the winter, the ungulates that live where there is too little food or too much snow migrate to other areas. Tigers migrate with the ungulates, but only to a certain extent. They cannot invade another tiger's territory. As a result, some tigers come down to the village for livestock, causing conflict with the villagers.

I left the Apasna Beach bunker and headed over to Petrova. The path along the coast was cold and rough, but the sea was calm in the low tide. In the light blue fog, seals were perched on every rock, enjoying the serene morning water. I climbed over the range by scaling a cliff. The sudden hike after a long and sedentary stake-out had me feverish and my legs shaking. I stood on top of the cliff and saw faraway Petrova Island in the dim morning light. Behind me were the jagged rocks of Apasna Beach. It had been a month since I'd moved to the bunker on the coast to get White Snow's family on film for the last time before the end of the winter stakeout, but I hadn't seen so much as a deer, let alone tigers.

I walked for several hours down to Petrova Beach. The mullet were running. A school the size of a grown man's forearm was gathered at the estuary, either because the freshwater was warmer, they were looking for food washed down by the freshwater, or they were there to feed on the sea lettuce that thrives in brackish water.

Stefanovich, who would normally be here during a low tide like this to catch trout with a spear, was nowhere to be seen. Chara, the four-eyed lodge dog who always came out to greet me, wasn't around either. The trout were jumping, but no one was there to see them. I crossed the beach and went into the lodge at the edge of the woods. There was nobody inside. I went back outside and found two footprints. The lattice pattern was from Stefanovich's boots. I followed his footsteps for about one hundred meters around the hill and approached two people looking at something near a spring at the foot of the hill. Zenya, the lodge keeper, saw me, slung his rifle over his shoulder, and drew a line across his neck with his thumb. Stefanovich gestured at me to come and see.

Tiger tracks were all around the spring. Most of them had been made the day before and had frozen overnight, but some of them were from that morning. The moisture gave the snowy, muddy prints—toes the shape of dates and the ball of the foot shaped like a pear—a glossy sheen. I pulled out my measuring tape. Nine and a half centimeters. It was the same size as White Snow's.

Stefanovich pointed at a spot about ten meters away from the spring. A spot in the leaves was pressed down. A tiger had been resting there and then chased after something. There was blood where the tiger's tracks ended. The blood was connected to the dog tracks coming from the lodge. Chara had tried to run, but spilled blood before she could get very far.

Zenya told us what had happened: "Around four in the morning, I woke up to the sound of Chara barking. She was barking pretty angrily and then let out a yelp, and everything grew quiet. I grabbed my rifle, went to the door and called Chara, but there was no answer. I thought there might be a tiger, but it was too dark and I didn't dare get out there on my own, so I fired three warning shots. I think the tiger had had its eye on the dog since yesterday. Chara was not herself. She was barking all evening. She barked all the way down around the corner where the spring is and then came back to the lodge. I should have figured it out sooner and fired some warning shots to shoo the tiger."

The tiger had taken the dog up the hill. We found tracks between the trees on the hill, alongside drops of blood. There was no sign of the dog being dragged, but it wouldn't have left traces anyway. She was so light and small. It was probably the gunshots that had sent the tiger deeper into the forest.

With Chara between its teeth, the tiger had met up with two other tigers in the oak forest about two kilometers away from the lodge. Their paw pad widths were 6.8 and 7.6 centimeters. Cubs. As I'd thought, it was White Snow who had hunted the dog to feed her cubs. She'd hid near the lodge all night and got her when she came close.

White Snow's cubs were a lot bigger than when I'd seen them three months earlier. We found traces of the two cuddling up and napping together. They'd waited here and gone out to greet their mother when she returned with the dog.

White Snow must have been very hungry to hunt Chara. There wasn't enough food in the forest to feed her cubs. The poor acorn yield in the area had brought on a dramatic shift in the food chain. Deer and boars had migrated to other areas in search of acorns, and the remaining ungulates were living on tree bark. To add insult to injury, more people were turning into poachers. There were no jobs, but there were mouths to feed. As well, professional poachers were increasing in number, and they brought with them tourist poachers from the city in greater numbers every year. The poaching was killing off whatever deer and boars remained. The Manchurian red deer had disappeared from the coastal range altogether, and it was getting increasingly difficult to find Ussuri sika and wild boar tracks. Life was dwindling in the once bountiful Ussuri forest.

White Snow would have learned from Bloody Mary that it's dangerous to go near manmade structures or livestock. But she was so hungry and had nothing to give her cubs that she remembered the brown-black dog she'd once seen when she came down to the lodge with her brother.

White Snow had fed the dog to the cubs on a hill tucked out of sight behind thick shrubs and oaks standing shoulder to shoulder. Chara's black and brown fur was scattered around. The cubs had left a web of tracks, too. White Snow's tracks showed she had not stayed long but had soon headed out on the mountain again. I assumed she had given the small dog to the cubs and then gone out in search of more prey.

The hungry cubs had eaten everything down to the little bones, leaving only the fur and the larger bones. But where was Chara's skull? We followed the cubs' tracks and found the skull thirty meters away. Chara had been such a small dog that there hadn't been enough to go around. One of the cubs had probably brought the head over here to have it to itself. The cub had eaten the scalp, too. The little bit of flesh remaining on the skull was still fresh. They had eaten Chara not long ago.

The cub who'd eaten the scalp had the smaller paw: the sister, Mapa. Kuchi Mapa's tracks were printed next to hers. It seemed he'd caught up with her after finishing the body. The two tracks mingled and then led to evidence of a skirmish. The mother's tracks were printed again over the signs of the cubs' row. Upon closer examination, I saw that White Snow's tracks were on top of all the tracks and traces her cubs had left. She must have joined them after all of this had happened.

We found a new trail of blood. It was a fresh trail, and not Chara's. Had White Snow brought them something else to eat after all? But the blood ran along the cubs' tracks, not White Snow's. There wasn't a single drop of blood on White Snow's tracks, only places where she had stepped on the blood. We followed the tracks.

"I don't believe this! How could this happen?" Stefanovich cried. There was a tiger cub lying on the ground. "What... where's the nose and feet?"

She looked like she was napping in the warm sun, but her stomach was gone and one of her hind legs had been torn off up to the rump. Her nose and one front paw were also badly injured. It was a gruesome death. The little cub who had jumped into Khajain's embrace and played rabbit with her father's paw was dead.

Mapa must have just died. Her body wasn't stiff yet. I looked up and around the forest. I felt a chill. I couldn't tell if the eerie energy was coming from the forest or from inside me. Who had killed this young cub? And why? I searched everywhere, but all I could see were the tracks of her mother and brother, who had apparently been here until moments earlier. It couldn't have been White Snow...

I saw tooth marks on the dead cub's neck. Tigers have powerful jaws, but not powerful enough to snap the neck of a large animal such as a wild boar or a bear, so they resort to strangling them instead. When hunting smaller animals, tigers snap the spine at the base of the animal's neck in one bite. If White Snow had killed her own cub, she would have broken her neck. But the cub's spine was clean, while the rest of her body was covered in cuts and bites.

The holes where four fangs had dug in were narrow and shallow, only four centimeters deep. A fully grown tiger's fangs are five centimeters long, but leave wounds ten centimeters deep. The bite marks on the cub belonged to another cub. There was a trace of another cub lying down in front of the dead one's body. Its paw was bigger than that of the dead cub. Kuchi Mapa had eaten Mapa.

The dog was the problem. One dog was not enough for the starving cubs. They must have got into a fight over who got to have Chara's head. They'd had a rough fight over the head and Kuchi Mapa had wound up killing his sister by strangling her. Still hungry after eating Chara's head, he'd started eating his sister's body. If their mother had brought them a ninety-kilogram deer, this would not have happened.

White Snow had arrived too late. She'd followed the trail of blood to find her son eating her daughter. Only when he'd seen his mother did Kuchi Mapa stop eating. White Snow had circled her daughter's body. This was a reality she had to face. There was sadness in her tracks—the agony of a mother who couldn't leave, but couldn't stay, either. This was the fate of a female tiger raising cubs in the dwindling Ussuri forest.

Stefanovich took off his coat and covered Mapa with it. He placed a cross made of sticks on top of her. "I'm sorry. Forgive us, little one," he said. "The forest used to be beautiful and full of things to eat, but there are so many problems now. No deer, no boars. Nothing for you to eat. There were five of you here just a year ago, and now it's just your brother and mother. Where did everyone go, huh? If things continue to get worse, there will be nothing left for us to protect anymore. We'll see what's left in five years . . . ten years . . ."

The old forest ranger shook with anger before this gruesome

reality. Once called the northern jungle, Ussuri was now a poachers' paradise. Dozens of Siberian tigers died every year, and there were only 350 left.

My knees buckled, and I sat. There was a hole in my stomach and the wind was blowing through it. I felt a raw, sharp pain from the anger of the old forest ranger and the grief of White Snow. Tears rolled down my face.

People would come out from the nature reserve office as early as the next day to investigate the scene. I looked at Mapa wrapped in Stefanovich's coat for a long time before I turned to leave. I walked away quietly and did not look back once. My gaze fixed on the ground, I suddenly stopped. White Snow's tracks overlapped the footprints we had made on our way here. White Snow, who had been near Mapa when we were heading toward them, had circled back through the forest and followed us. When we had found Mapa and were examining the surroundings, she had been nearby, watching us. Now that we had left, she was probably back by Mapa's side.

I turned around. In the morning light, the oaks on the hillside cast uniform shadows on the ground. The shadows quivered in the wind and whistled like electric lines. The yellow leaves that had managed to hang on through the winter fluttered. White Snow was in that oak grove. She was pacing and sniffing her cub wrapped in a coat. We hurried on so she could grieve in peace.

We returned the next day for the official investigation of the scene. The cause of death, the officials decided, was strangulation by a sharp fang. The scene was exactly as it had been the day before, except the coat had been peeled off. White Snow seemed to have taken the coat off Mapa to see what we'd done to her or to look at her cub one last time. And then she'd headed for the coastal range with Kuchi Mapa.

The people from the nature reserve abandoned the idea of

taking the tiger's body with them. Their original plan was to preserve it as taxidermy, but the body was too damaged. To prevent black market trading, we decided to cremate her.

That evening, we invited the ginseng-gatherer couple from Mayak Village to a small ritual for Mapa. For these people, the tiger is Amba, the spirit of the forest. The sight of tiger tracks in the forest comforts them, and they mourn the death of tigers. But they also believe that death is the beginning of a new life, not the irrevocable end of one. We needed a ritual to lead the young cub to a new life.

We gathered dry branches from the forest, stacked them into a pile, and lay Mapa on top. We lit the fire. A pungent smoke rose and the fire burned quickly thanks to the tree sap. Aktanka sounded the drum. Olga Kimonko, wearing traditional dress, danced the shaman ritual dance around the fire to the rhythm of the drum. I didn't understand the words to the song she sang, but it was heartbreaking. The Udege had held this healing ritual since ancient times to send the dead to the underworld and Amba to the world of the living. We held this ritual that day in honor of all things disappearing from this forest.

The smell of burning flesh spread. The slow movements of an old shaman flickered on the other side of the flames. In the fire, White Snow's daughter burned and turned to ashes. A young soul was departing. Bloody Mary was gone. White Sky was gone. The bravest, most sacred species was dying at the hands of man. As the flames rose, so did Olga's song. Aktanka turned the drum over to his son. Ado had gone into the city in search of wealth you can only dream of in the forest, and in the end chose to return to the forest. Everyone at the ritual prayed that Ado would one day hand over the drum to someone else who would keep it beating, and that another life would fill the empty space that this young life had left behind. For the sake of Ado's future and the future of the forest, and the future of Amba who protects the forest, Olga poured her heart and soul into her prayer until the flames died.

After the ritual, Olga said to her son, "Do you know how great the forest is, Ado? There are so many trees and animals. Amba rules the forest. People can cut down trees if they want. The forest will come back to life again. Amba, too."

CHAPTER 31

The Grief of the Living

THE TIDES ROLLED in and out, crashing on the shore and retreating like any other day. The sea was quiet at low tide. Birds found their way home in the fading light and dissolved into the horizon, where an island floated in the distance.

Green hillocks extended down the coastal range and continued onto the coastal basin. The hills looked like a row of gravestones. The blue evening light on the gravestones created the illusion that the slumbering soul of Sikhote-Alin was waking up. A head appeared above one of the mounds I happened to be blankly staring at. The head bobbed as the creature came over the hill. It was far away, but it was definitely a tiger. The tiger's orange fur was tinted blue in the evening light as it came over the hill. A smaller tiger followed the first. The small tiger was limping—almost hopping—along. The big tiger waited for the little tiger to catch up, and the little one limped past it. The gap between the two tigers closed and widened as they disappeared over the hill.

The mountain shadows dissolved in the ambient light, and the blue evening turned to night. A bright, full moon rose above the mountain range. Diplyak Basin was dark in spite of the moonlight.

The basin gets a lot of sun during the day, so the snow was almost all gone. The white expanses of snow that had reflected and amplified the light were now gone, and the leaves on the forest floor absorbed the moonlight and whisked away the moon's energy.

The mountain path the two tigers had climbed led to this basin. I swapped in the night lens and waited. The wind was calm and the air clean and cool.

Around eight o'clock, a shadow even darker than the forest came into the shrub-lined basin. The shadow took great strides forward and then stopped. Its gold, twinkling eyes carefully scanned its surroundings. The cloud blocked out the moon, and it became even darker. The shadow waved its long tail, turned around, and called at the forest behind it.

The cry traveled from deep within along swelling vocal cords and sent its vibration through everything in the basin. A faint sound responded in the forest, and a small shadow appeared. It was limping. Each time it limped, it jerked its tail to keep its balance. It was Kuchi Mapa, the brother who had survived the fight for Chara's head. It appeared he had hurt his leg during the fight.

The limping cub walked up to his mother, nuzzled her from her waist up to her chin, and whimpered like a child. He may have been hurting or hungry; either way, he didn't seem happy. White Snow rubbed her cheek against his and tried to console him with the same nasal, whiney groan he'd made.

White Snow always brought her cub to this pine grove when she was in the area. The space under the drooping pine branches was cozy, like the spot under a parasol. White Snow lay down on the soft bed made from decades' worth of pine needles. On one side of the grove lay the white skull of Five-Prong, the deer Khajain had caught for this family. The five-pronged antlers were just as majestic now as they'd been when they were attached to the deer that had stood with its hooves underwater, eating seaweed. Kuchi Mapa nuzzled into his mother, and White Snow licked him meticulously. He licked his injured leg. Each time his whining got louder, White Snow made a rolling, guttural sound to soothe him. The gentle snuffles of the mother and son sounded like the whispers of forest spirits. A fat raccoon waddling by the bushes turned toward the sound. It belatedly recognized the source and bolted into the bushes. The nasal sounds gradually let up, and Kuchi Mapa fell asleep with his head on his mother's stomach. White Snow also closed her eyes with her chin on her paw. Neither one budged, apart from the rising and falling of their chests as they slept. White Snow had lost one cub, but it was back to life as usual.

The night deepened, and the moon sailed to the top of the sky. The tide rolled back in and started lapping at the shore. The slumbering winds were now awake. White Snow also awoke from her light sleep and turned her head back and forth to take in the scent of the pine grove. Kuchi Mapa woke up to the sound of his mother's movement. He scratched his chin with his uninjured hind leg. His cute face still looked sleepy.

The mother teased her son. The whispers of the forest spirits resumed. To get down to his eye level, White Snow lay on her back; then she gently hit Kuchi Mapa in the face with her front paws. She rubbed her forehead against his face and then bit him softly with her big fangs. Acting silly, Kuchi Mapa opened his mouth to nip her back. White Snow played with him, controlling her strength so as not to hurt him. The mother had taken over the role of the younger sister.

Kuchi Mapa soon lost interest. He must have been tired after a day of limping behind his mother through the woods. He kept licking his injured leg. I couldn't tell if he had injured the bone or a tendon, but he could hardly put any weight on it. If he was permanently crippled, it wouldn't be easy for him to survive in this unmerciful game of life. His mother was around to look after him now, but when the time came for him to leave his mother, or when his mother's instinct to mate pulled her away from her cub, survival would be tough for him. If he couldn't survive the itinerant, hunting lifestyle, he might himself end up a white skull rolling in a pine grove in the near future.

White Snow picked herself up and yawned widely. She stretched her spine by extending her front legs out straight and pushing back. Her handsome tail was raised high. Now ready to leave, White Snow grunted and nudged Kuchi Mapa in the rump with her snout. She headed down the path. He limped after her, then stopped. She walked for a while before turning to see why he wasn't following. Kuchi Mapa sat and licked his injured leg. White Snow stopped again and gazed at him. Her golden eyes looked empty, as if she'd been hollowed out.

She hesitated, then came back to nudge him again. He pawed her hind leg and bit her tail, begging to play instead of wandering through the forest. She turned around and walked away. He watched her go, his head still poised to bite the tail that had already slipped out of reach. He didn't move in the slightest, taken aback that she wouldn't play with him. Once she started to dissolve into the shadows, the cub limped after her, his tail jerking up with every limp. The bobbing tail reflected the moonlight all the way into the dark forest. It was the burden of the living to journey on.

CHAPTER 32

The Tiger Lives On

THE FROZEN DEER River glistened in the morning light. I saw a thread or two of spring spiderwebs catching the sunlight as they fluttered over the ice. I heard water flowing. The river was thawing itself from the inside out and was beginning its flow by pushing its way through the little cracks between the ice and rocks. Moisture rose from the river, mingled with the breath of trees, and hung on the tips of branches like sheer gloves. The air was clear above the frosty mist, and the sound of birdsong and the quickening, subtle pulse of life flitted around the forest. I felt the stillness of the forest and, within it, the movement of life.

Creek. Creek.

Something was treading on the brittle ice. A tiger came out from behind a cliff and slowly walked down the middle of the frozen river. Its footsteps on the ice had an innate weightiness and the languidness of someone who has had a long night. The tiger must have been returning from a meal, perhaps a deer—speckles of blood clung to its face. Its bones were strong, its muscles firm, and the black and orange fur was clean and glossy. The old trees on either side of the river were its guards, and the birds flitting from tree to tree pecking at worms sung its praises.

The tiger turned around to look at the upstream cliff it had just come from. From a spot where a large tree cast a shadow on the silver ice, a little tiger made a careful entrance. Behind the little tiger, a second mischievous face appeared from behind the black cliff and sauntered out carelessly. Hansel and Gretel looked as radiant and lively this morning as if they were spirits of the forest themselves. The appearance of White Moon and her cubs quickened the pulse of the river, and the heart of the forest began to race.

White Moon looked at the end of the winding Deer River, the very spot where, one hundred meters downstream, my bunker was dug into the foot of a hill. Pale pink flowers nicknamed "roe deer ears" had bloomed next to Amur Adonises. Sunlight had melted the snow. It shone on the dry leaves and blades of grass, and the yellow flowers blooming among them each had a little bit of spring in them.

I saw flashes of anxiety in White Moon's tired eyes as she gazed at either the bunker or the Amur Adonises—I couldn't tell which. The cubs came to her side, careful not to step on the melted ice, and stared in my direction apathetically. The bunker was camouflaged, buried under layers of leaves, melted snow, and spring flowers. It had been disintegrating underground and turning into a part of nature for quite some time, so I was sure I couldn't have been better camouflaged. The wind was gentle, so there wasn't much chance it would carry my scent to her. Still, White Moon must have picked up my scent somehow, because she appeared to suspect that something was going on in the spot where I had melted into the forest. Fortunately, her gaze wasn't very tense and her suspicions were vague. Drops of melted snow dripped from the tree towering over the bunker.

White Moon turned away and lay on the ice. The cubs did the same. White Moon rolled on her back, her white belly fur now facing up, and shuddered as she stretched her legs to the sky. The cubs also rolled on the ice with their backs against each other. The whole family enjoyed the morning in the warm sun.

Now six to seven months old, White Moon's cubs were healthy and grown up enough to handle a spiteful badger and stalk a fawn. Hansel was shedding his mischievous childlike face and turning into a handsome youth. White Moon had raised her cubs well, thanks to the abundance of acorns, nut pines, and walnuts in Deer Valley the previous fall, along with her careful personality, which reminded me of Bloody Mary.

Gretel managed to pull out a frog from the hole in the ice. Still not quite awake from its winter hibernation, the frog hopped on the ice twice and then stopped. Gretel nudged it with her paw, trying to get it to hop again. When it moved once more, she stalked it like a cat stalks a half-dead mouse. Hansel saw this and casually walked over to her. He pretended he didn't care, then suddenly popped the frog in his mouth. He must have been jealous of his sister's long-legged hopping toy or curious what the slick, hairless creature would taste like. Gretel stared at his mouth. He chewed and swallowed. She turned and walked off.

Gretel wandered up the riverbank strewn with yellow Amur Adonises and smelled each flower. She found a nice cluster of them, plopped down in front of it, and put one in her mouth. She moved her face and jaw muscles and chewed so intently it looked like she was eating delicious boar meat. Hansel saw this and rushed over to the riverbank to eat a flower, too. Seconds later, he made a face, spat out the flower, convulsed in dry heaves, and finally spat something up. He licked the top of his nose as though he was trying to get the awful taste out of his mouth.

Hansel found a strange-looking mushroom and decided to check it out. He became immersed in alternately putting the mushroom in his mouth and spitting it out, experimenting with nature. The brother and sister were discovering the world under the care of their mother. The sun was warm, and the Amur Adonises on the riverbank glittered like gold.

Hansel lost interest in the mushroom and conspired with his sister. The two carefully approached their mother and pounced on her at the same time. White Moon, of course, saw this coming. She lay on her back and met the cubs' challenge with her legs in the air. Hansel clamped down on her forehead, and Gretel climbed on her back. When the cubs started to get to her, White Moon got up. Hansel jabbed her in the head a few times like a boxer, and White Moon swung her paw back as if she was going to hit him hard, but just shoved him instead. Gretel climbed on top of her mother like she was riding a horse, and White Moon responded by pretending to bite her in the head, but just licked her. Hansel used this opportunity to bite White Moon in the neck. White Moon opened her mouth wide and grabbed his whole head in her mouth as though she was about to swallow it whole. To deal with the little one from the back and the big one from the front, she lay belly-up on the ice again.

Even as White Moon wrestled her cubs, she drew a boundary between the world and the safe psychological sphere where her family was. Outside was where she had to remain levelheaded and keep the world at bay, but inside was a warm place for bonding with her family. She was vigilant and prudent outside, but as innocent as a puppy inside. The cubs were at ease, and their mother was affectionate. They were a real family who shared mutual trust and dependence. People think of tigers as terrifying, fearless creatures, but they probably spend most of their lives in the peaceful, affectionate state I saw them in now.

But people want to see tigers doing sensational, tough things. They don't think it's very exciting to see tigers rolling in the snow with their bellies exposed. But for me, watching White Moon's family with my own eyes was the most exciting moment in my life with them. You get to see a wild tiger frolic with her cubs only when you've managed to reach the most intimate, innermost part of nature. Only when she feels that safe does a mother tiger reveal glimpses of her intimate family life. At that moment, I was truly and completely one with nature.

If I wanted to see them do something sensational, I could just step out of the bunker. This would prompt White Moon to attack me with the fury associated with tigers or reveal her beastly side and leave. I could capture this brief but thrilling moment on film and show it to people who didn't understand her life but wanted to see something wild. But should I knock down the boundary she had built between her family and the outside world? Should I shatter the bubble and disrupt her family's peace, all for some footage?

Although White Moon didn't know I existed, I had watched her grow, and every one of her gestures and experiences had inspired such strong emotions in me. I had wondered about her when she'd chased Manchurian trout in Deer River with Bloody Mary and when she'd left a long, clean stretch of tracks on the sandy beach. I had pictured what her face looked like, constantly drawing and erasing her image in my head with increasing curiosity. When she'd sat primly on the snowy hill and looked down at the blue East Sea, I'd pictured her as the tiger equivalent of a girl in an orange dress sitting on white silk, and when she'd attacked the bunker with her siblings, I had felt her every breath as clearly as if she were right up against my ear. Every gesture and every trace of her made me feel alive. All these encounters had led to this moment when I got to watch her play. White Moon, her family, and I may not have shared an obvious connection, but we were linked in ways we couldn't imagine.

White Moon was as relentlessly careful as her mother had been, and her affection for her cubs was as sincere as the Ussuri forest.

Looking at her, I knew that nobody had the right to violate the peace of this family.

True to their silly nature, the cubs wouldn't leave their mother alone. They were going through a growth spurt and had to do everything and anything to expend their energy and discover what new things they could do with their growing bodies. The games they played were a rehearsal for surviving in the wild. Nature is a battlefield and their family members were sparring partners. The two-against-one struggle was getting tougher. In Korea, they say if you have a tiger dream when you're expecting a child, you'll have a child who looks after its parents. Based on what I had observed, I'd say a tiger child would mean endless mischief.

White Moon couldn't take it anymore. She got to her feet and walked upstream. Gretel reluctantly followed. Hansel lay flat on his stomach, holding the ready-to-pounce position he was in when she walked off, shook his head hard, and got up as well. White Moon's family receded into the icy landscape, leaving behind long tracks of floral paw prints on the ice.

Sunlight beckoned spring as it shattered across the empty, silver river. A film of frosty mist hovered in the dense forest of naked trees on the other side. This was once the Ussuri jungle, home of the Siberian tiger. No one called it that now. But the tigers lived on. I prayed with all my heart that, like White Moon's mother, Bloody Mary, and her mother before her, Hansel and Gretel would also have children here and bring them up safely.

EPILOGUE

Diamond Dust

SPRING WAS OFF at a sprint before I knew it. The new warmth brought the azaleas, which were followed by foxtails on the riverbank. Leaves and shoots painted the land green, and flowers drank in the morning dew that fell in the spring showers. When summer came, it was hot and muggy. Deciduous tree leaves hung limp in the heat, and the humid air clung to the skin. The moisture rising from the heat of the earth slithered and licked along the mountain ridge like transparent flames. The mountains panted like a beast that had just chased a deer.

The leaves that glistened all through the summer heat turned color again, and pools of light formed wherever the leaves fell. Even in the cool autumn light, the exposed underbelly of the forest was dry like hay. Leaves rolled and rustled on the ridges and in mountain valleys and whispered secrets to each other all day. Like a beast putting its cold paws together and covering them with its fluffy tail, the forest covered itself with a warm blanket of leaves and prepared for its winter sleep. Snow fell on everything under heaven and covered the mountains like a blanket, and the forest drifted off for a long nap.

The frozen, snowy Kievka River snaked down its winding course. There were fresh floral paw prints crossing the river. The moisture in these tracks had frozen, giving the prints a glossy sheen. There was even a sharp line between the snow and the prints. The prints had been made the previous night, after a dusting. There was a groove like the mark of a stick dragging in the snow where the left hind paw landed. The tiger was dragging its left hind paw behind it. The front paw pad width was 10.7 centimeters. It was Kuchi Mapa. No blood. The winter sun fell across the path of Kuchi Mapa, who had limped across the snowy Kievka River alone. The frail life of an animal on the precipice was palpable in these tracks.

The limping prints led to the site of the accident. Kievka River flows past downtown Lazo, continues south to Kievka Village, and heads into the East Sea. A new, unpaved road ran for eighty kilometers along the river between Lazo and Kievka Village. The accident had taken place on that road, not far from Kievka Village.

The tracks in the thin strip of willows by the river were apprehensive. The limping must have held Kuchi Mapa back for a while. When he had finally jumped into the road, he'd been hit by something and landed in the middle of the lane. There was a pool of blood where he'd landed. Then he'd gotten up and staggered into the forest.

The road was quite wide. Kuchi Mapa had probably stood on the side trying to decide whether or not to cross when he'd panicked at the strong headlights speeding toward him and jumped into the road. When something speeds toward animals in the cat family, they try to outrun it instead of waiting for it to pass. This comes from an instinctive trust in their speed and agility. His faith in himself, in spite of his injured leg, had led to the tragedy.

The driver told us that something orange had suddenly jumped in front of the headlights. I'll bet Kuchi Mapa ran as fast as he could in spite of his leg, but he could not have outrun the land cruiser jeep speeding down Kievka Hill. The driver said the animal had hit the bumper and sprang outside the pool of his headlights.

When the driver turned around and shone his headlights on the road, the animal had already gone. He said he couldn't see very well, but it looked like a tiger. There was a big dent in the front bumper of his jeep. The impact would have been considerable.

The tracks continued through the fir grove on the other side of the road, scrawling an *S* on the ground. Kuchi Mapa's stride became narrower, and the back leg dragged more the farther the tracks went. The front paws started to drag as well, which was something that only old male tigers did. There were drops of blood in the snow. But what got to me was how the tracks suggested he had been shaking with every step he took.

We followed the tracks for about two kilometers and came to the spot where he had rested. In the snow was a deep imprint of a tiger from head to tail. There was frozen blood in the stomach area, and the amount of blood suggested he had lain there for quite some time.

A fresh set of prints lurked around Kuchi Mapa's imprint in the snow. The front paw width was nine and a half centimeters. It was White Snow. White Snow had caught up with her son after being off somewhere on her own. She must have been stalking prey. There was a depression in the snow of her where she had lain next to him. Comprehending the situation he was in, she had lain by his side and licked his wound.

Kuchi Mapa must have regained his strength after some time. The tracks continued as he'd walked with his mother. There was another indentation in the snow alongside Kuchi Mapa's footsteps. It looked like a stick had been pulled through the snow. He'd been dragging his tail. Tigers never so much as let their tails touch the ground when they walk. Unless they're trying to change direction in the middle of sprinting or they're walking through snow that comes up to their knees, their tails never drag. The only time their tails hang that low is when they're about to die.

Half a kilometer farther in, we came to a cozy clearing surrounded by firs. Big trees towered over the younger, shorter ones. Some trees that had lost the battle for sunlight and nutrients looked on quietly, dry and dead. The northwesterly wind traveling over the mountain range shook the fir branches. The snow on the branches blew in the wind. The diamond dust that had been hanging in the air since morning mingled with snowflakes and fell down as glitter. Here, White Snow's son lay frozen in eternal sleep. He had won his battle with his sister, but could not prevail over his own fate.

Kuchi Mapa was thin and bigger than when I'd last seen him, but smaller than other cubs his age. He had broken several ribs in the accident and bled internally. Just like when his grandmother, Bloody Mary, had died, his side had split open and his intestines had spilled out. The blood was already congealed and frozen. His injured leg had healed, at least on the surface, leaving old scar tissue on the shin. Eighteen months old. If he hadn't hurt his leg, he would have been semi-independent by now. And alive.

White Snow had paced as she'd waited for him to get up. She had tried to make him get back on his feet and continue on. There were deep pits in the snow near her son's tail and back where a snout had dug under the body and tried to push it up. She had known he was dying. She'd also known there was nothing she could do. Marks of White Snow's despair were carved into the snow. I wondered what the son had thought as his mother kept nudging him. In the end, he wasn't able to get up. The life of White Snow's son, at least in this world, came to an end. Perhaps it had all been fate. White Snow was her son's one chance in life. He wouldn't have been able to make it on his own, but would have been forced to anyway. He had come to the end of his perilous life in this cold snowy bed.

His eyes were shining. He didn't look like he had suffered. In his eyes, I saw the empty space after the soul has departed, entrusting its body to nature. I closed them. I dusted off the snowflakes that clung tenaciously to his stiff body. We carried his body onto a sled, took it to the road where a truck waited, and transported him to the nature reserve office. White Snow's son would be taken apart and taxidermied.

Three days later, I visited the fir grove again. There were fresh prints where we had found White Snow's son. White Snow had still been there. She had been lingering in the forest where he'd died, looking for him, or missing him.

A dusting of snow came down the next day. Snow powder descended slowly from the sky from early morning on. All traces of the past were buried under snow, and nature was ready to record a new chapter. There were no more flower-shaped paw prints. I didn't know how White Snow would remember this time in her life, but she had left without a trace to erase her past and write her life anew.

Cold morning light filtered in through the jagged fir branches. A rainbow appeared above the powdery snow, which floated down like dust. Loose piles of white on the fir branches blew away with needles in the gentle breeze. The fresh snow and the diamond dust in the sun turned the forest into a sparkling, magical place, and the sunlight shone in all the colors of the rainbow.

ACKNOWLEDGMENTS

THANK YOU TO my family, my friend Stefanovich, Kalesnikov, Dr. Galina Salkina, Mike Birkhead, Jay Eberts, and all the people who helped with the publication of this book—including my translator, Jamie Chang, publisher Rob Sanders, editor Jennifer Croll, and everyone else at Greystone Books.

THE SIBERIAN TIGER PROTECTION SOCIETY

In 1995, Sooyong Park, alongside Dr. Galina Salkina and Vladimir Kalesnikov from Lazo Nature Reserve, established the Siberian Tiger Protection Society (STPS) to protect and study wild Siberian tigers. Every year the STPS works to reduce poaching and trapping, prevents avoidable tiger deaths in villages, educates local villagers about tigers, and studies wild tigers in the field.

For further information on the STPS or to find out how to support our work, please write to:

SIBERIAN TIGER PROTECTION SOCIETY
111-401 Garam, 127 ILwon-ro, Gangnam-gu
Seoul, Korea
tiger.hut2.ru
tigre1218@naver.com
Tel: 82-10-7302-1928

INDEX